CREATION

versus

EVOLUTION

Scientific And Religious Considerations

Arlo E. Moehlenpah, D.Sc.

Arlo E. Moehlenpah

4/20/02

Creation versus Evolution: Scientific and Religious Considerations

Copyright © 1998
Arlo E. Moehlenpah

Published by: Doing Good Ministries
                 8632 Blue Grass Drive.
                 Stockton, CA 95210
                 (209) 957-6605
                 www.DoingGood.org
                 E-mail: Moehlenpah@aol.com

Cover design by Paul Povolni

Cataloging-in-Publication Data
Moehlenpah, Arlo E., 1936-
        Creation vs. evolution: scientific and religious considerations
        Includes bibliographical references and indexes
        ISBN 0-9667054-0-8
        1. Creation. 2. Evolution–Religious aspects–Controversial literature.
        3. Bible and evolution. 4. Religion and science. I. Title.
BS659.M64 1998   231.7'65–dc21

Table of Contents

Acknowledgements

This book is the culmination of thirteen years of teaching geology at Christian Life College in Stockton, California. I thank God for the opportunity to teach this class and for the students who contributed in many ways including thought provoking questions. Most of what I know I have learned from others. I have tried to carefully document the sources of my information. I apologize for any errors or omissions. I especially want to acknowledge the contributions of Dr. Henry Morris and the staff at the Institute for Creation Research for their writing and seminars, which have had a great influence on my thinking.

In addition, I want to thank my friends, Ron Schoolcraft, Brenda Leaman and Nancy Hunt for their help in proofreading the book and making suggestions for improvements. I thank my wife, Jane, for her typing, editing and computer skills that aided greatly in the production of this book.

I also want to thank God for the physical and mental strength that He has given us to accomplish this project.

The photos used in chapters 1, 8, 16, 31 and 34 were obtained from the Microsoft Office 97 Clip Gallery 3.0 pictures.

The images used in chapters 5, 6 and 12 were obtained from the Microsoft Office 97 Clip Gallery 3.0 Clipart.

The images used in chapters 3, 7, 10, 14, 15, 17, 19, 20, 27 and 32 were obtained from IMSI's MasterClips Collection, 1895 Francisco Blvd. East, San Rafael, CA 94901-USA.

The photos used in chapters 23, 28, 29 and 38 were obtained from Expert Photo CD Gallery #2, Expert Software Inc., 800 Douglas Rd., Coral Gables, FL 33134.

The following table gives the URL's for photos taken from the internet.

Chapter	URL
9	http://vger.rutgers.edu/~tempest/marsup2.htm
13	http://zygote.swarthmore.edu/evo5.html
18	http://www.pica.army.mil/pica/about/blasts/blasts4.html
21	http://www.lhl.lib.mo.us/pubserv/hos/dino/owe1863.htm
22	http://www.dinofish.com/image13.htm
24	http://vulcan.wr.usgs.gov/ljt_slideset.html
26	http://www.swcp.com/amha/dentranch/Text/dentsales.html
29	http://www.unlitter.com/sandcastle/anisand.html
33	http://marsweb.jpl.nasa.gov/

Preface

What the book is about.

This book explains what true science is and how science benefits mankind. It shows that nothing truly scientific contradicts the Bible but the theory of evolution, which is falsely called science, contradicts not only the Bible but scientific laws and mathematical probabilities. This book refutes many of the popular evolutionary ideas from the fields of anatomy, astronomy, biology, chemistry, dendrochronology, geology, paleontology, physical anthropology and physics. Evidences for the Biblical account of creation and the Flood are clearly presented.

Some other interesting topics covered are dinosaurs, origins, animal migration, the origin of races and the geographic distribution of animals.

Format of the book.

This book is written in a question and answer format. After reading the first chapter the reader need not follow the exact order of the book but instead may choose topics that are of most interest at the time.

Purpose of the book.

The educational systems and museums regularly bombard the public with evolutionary concepts of origins and the age of the earth. Despite their claims, most of these ideas have absolutely no scientific basis. The purpose of this book is to prepare the reader to recognize what is truly scientific and what is falsely called science and to help the reader see that none of the Biblical account of creation and the Flood needs compromising. What we actually observe scientifically fits in well with the Bible record. This book will give the reader answers to many of the questions that are asked regarding origins and will help strengthen the faith of those who are struggling between the two opposing views of origins.

Intended Audience.

Any Christian who reads or listens to the media regarding evolutionary concepts will profit from this book. I am hoping also that this book will be of evangelical benefit. It has been said that the theory of evolution has caused more people to lose faith in God and the Bible than any other idea ever proposed. If we cannot get people to believe in the God of Creation we will not get them to believe in the God of Salvation.

My experience in teaching on these topics both in Bible colleges and in our churches shows that there is a great interest in this subject and that people want answers to the questions they are asking and being asked.

I also sincerely hope that the concepts of this book will get into the minds and hearts of every Christian especially those who plan to attend secular college. This book also can be used in Christian schools and Bible colleges as a text for a course that might be entitled "The Bible and Science."

1 / Science vs. Science Falsely So-Called

How is the word "science" used in the Bible?

The word "science" is mentioned twice in the King James Version of the Bible. During the time of the exile, Nebuchadnezzar, king of Babylon, decided to teach the learning and tongue of the Chaldeans to a select group of the children of Israel. Some of the qualifications that they had to have to be selected were that they must be "skilful in all wisdom, and cunning in knowledge, and understanding science" (Daniel 1:4). Among those chosen were Daniel, Shadrach, Meshach, and Abednego. From the context, "understanding science" was a very positive quality. The second time the word "science" is used in the Bible is in I Timothy 6:20, "O Timothy, keep that which is committed to thy trust, avoiding profane and vain babblings, and oppositions of science falsely so called. Which some professing have erred concerning the faith." Paul closes this epistle strongly warning Timothy about "science falsely so called." Note that Paul does not warn against "science" but against "science falsely so called."

How is "science" defined?

It is important that we know what science is and are able to distinguish between "science" and what is "falsely called science." A typical definition of science is that it is a branch of study concerned with observation and classification of facts, especially with the establishment of verifiable general laws, chiefly by induction and hypothesis. Webster defines science as "systematized knowledge derived  from observation, study, and experimentation..."[1] You can look at various dictionaries and get slightly different definitions but the key words will be "observation," "experimentation," "verifiable," "testable," and "repeatable." In other words, if it cannot be observed, repeated, verified or subject to experimentation, then it is not scientific. Keep in mind that there may be a difference between what scientists actually observe in contrast to the opinion of scientists. For example, scientists may observe and determine what the liver does but that is different from how the liver originated.

What are the steps in the scientific method?

For years I have demonstrated the steps in the scientific method by the following pendulum experiment:

1. *Problem or Question.* What effect does the length of a pendulum have on the period of a pendulum? The length is the radius of swing from the pivot point to the center of mass at the bottom. The period is the time for the pendulum to swing back and forth.

2. *Hypothesis.* This is what I call an educated guess as to the solution to the problem. It is something capable of being investigated, hopefully to be proved true or false. Some possible hypotheses are: 1) the length does not affect the period; 2) as length increases, the period increases; 3) as length increases, the period decreases. If students choose hypothesis 2, I try to get them to be more specific such as, the period is directly proportional to the length, i.e., if the length is doubled, the period is doubled. A possibility for hypothesis 3 is that the period is inversely proportional to the length, i.e., if the length is doubled the period will be cut in half.

3. *Controlled Experimentation.* This is where the scientific method varies from other methods of solving problems. To perform this experiment all one needs is a fishing line with a sinker attached to the end, a ruler, a stopwatch, and perhaps a ladder. Tie one end of the fishing line to a cup hook attached to the ceiling and measure and record the length from the cup hook to the sinker. One person pulls the sinker to the side and lets go. Another person times the swing back and forth. To get more accurate results, time ten swings and divide this time by ten. Repeat several times and then change lengths. Do this for a number of different lengths.

4. *Correlation of data into a generalization or law.* This can be done by arranging your data in order of increasing length in two columns, length and period. This data should show that as the length increases the period increases. A better way to visually correlate the data is to plot the data on a graph with the length on the horizontal axis and the period on the vertical axis. You will see from this that although the period increases with length it does not increase proportionately with length. If, however, you plot the period versus the square root of the length it will be a straight line showing that the period of a pendulum is directly proportional to the square root of the length.

5. *Check the law.* Remember, if the law you arrive at cannot be verified or repeated, it is not scientific. Others must be able to do the same experiment and come up with the same results.

A simpler way to demonstrate the scientific method is with three separate balls of approximately the same size but of different masses, such as a golf ball, a ping pong ball, and a plastic "wiffle" ball. A plastic practice golf ball will also work. After seeing that they all have different weights, state the problem or question, "Which ball will fall the fastest?" You might guess one of the following: 1) The golf ball because it's the heaviest; or 2) the ping pong ball because its the smoothest; or 3) the wiffle ball because it has holes in it; or 4) they will all fall at the same speed. These guesses are known as hypotheses.

What are some ways to solve problems other than the scientific method?

After the hypothesis step the scientific method differs from other methods. In the "democratic" method, for example, you would take a vote and the hypothesis that has the majority of votes would be the winner. The whole process is over, but the majority vote may be incorrect. In both the pendulum and ball dropping experiments the majority of people usually vote incorrectly. In the "appeal to higher authority method" you try to find out what some authority has to say on the subject and what he says is final. For years opinions of "higher" authorities such as Aristotle were held as correct until proven wrong by scientific experimentation.

Galileo, it is told, dropped two balls of different masses from the leaning tower of Pisa. What did he find in his scientific experiment? What he found is a matter of history, but if it is scientific you must be able to verify his findings. Whereas you may not be able to go to Italy, you can drop all three balls from a stepladder and have another person see which ball hits the ground first. Is your curiosity aroused? Try the experiment. Isn't science fun? What did you observe? If you didn't observe it, or if no one else observed it, or if it cannot be observed, then it is not scientific but rather falsely called science.

What are some questions that should not be answered scientifically?

The scientific method is not the best method to solve all questions or problems. For questions such as "Who should be president?" we certainly don't want to experiment and let every person have a chance at being president of the United States of America for a period of time. This would be a sure way to usher in World War III or a revolution! Instead, we use the democratic method and have an election. For questions regarding morals such as "Should I smoke cigarettes, do drugs, drink alcohol, or engage in premarital sex?" it is wise to appeal to the higher authority of parents and the Bible rather than experimenting and ending up a physical and moral shipwreck. Likewise questions regarding direction such as "Should I get married, go to college, or join the armed services?" are better answered by appealing to higher authority. If we have questions about the past we can best find the answers from reliable

3

historians. By this I mean either people who were there or else those who heard from people who were there. There definitely are problems and questions which must be solved by other means or ways than the scientific method.

What is the difference between a theory and a law?

A *theory* is an explanation, thought or fancy as opposed to fact or practice. An example of a theory is the atomic theory. In reality both evolution and creation are theories in that neither is observable nor repeatable. Theories are ideas, which usually try to explain how or why, but are usually not observable. A *law* is a statement of a relationship or sequence of phenomena which is invariable under the same conditions. An example of a law is the law of gravity. Laws like this can be observed to happen over and over again, whether or not we can explain why. All scientific laws and theories, however, are subject to modifications with the accumulation of more data and the increase of knowledge. The law of conservation of matter stated that matter could neither be created nor destroyed, but when Einstein recognized that mass could be changed into energy the law had to be modified to be the law of conservation of matter and energy. Science does not claim for itself absolute truth. The Scriptures, though, are absolute truth while science is, at best, relative truth.

Why can't questions regarding origins be answered scientifically?

It is impossible to design scientific experiments to answer questions such as "How did life, man, the earth or the universe originate?" Since we can't observe, repeat, or verify origins, the scientific method does not apply. To find the answer to the questions regarding origins we must go to the word of the One who was there. The only One who was there was God. The question God asked Job, "Where were you when I laid the foundations of the earth? Tell me if you have understanding" (Job 38:4 NKJV), should be asked to proponents of evolutionary theories of origin. Were you there? Did you observe it? How do you know?

Is the Bible scientifically accurate?

Whereas the Bible was not intended as a textbook of science, the statements in the Bible when properly understood are scientifically accurate. One amazing example of the accuracy of the scriptures is found in Psalm 8:8, "...And the fish of the sea that pass through the paths of the seas." Matthew Maury, who was in charge of the depot of charts and instruments in the hydrographic office of the United States Navy from 1841 to 1861, recognized the significance of this verse along with Ecclesiastes 1:6 which describes the wind circuits. He realized that if he could plot the wind circuits and the ocean

4

currents it would be of tremendous value to the sailors of that day. By doing this he reduced the time required to cross the ocean by as much as three weeks.[2, 3] Another scripture that has been confirmed is Job 38:16, "Have you entered the springs of the sea? Or have you walked in search of the depths?" Job had no experience with undersea springs. The ocean is very deep and most of the ocean floor is in total darkness. It was not until the 1960's that scientists, with scuba diving equipment, discovered hot springs in shallow water near the coast of Baja, California. In the 1970's scientists, using deep diving submarines, located hot springs in the Galapagos Rift in the Pacific Ocean. Although only a small portion of the ocean floor has been examined it has been estimated that forty cubic miles of water flow out of the earth's oceanic springs each year. This discovery of oceanic springs was considered one of the foremost scientific accomplishments during 1970-1979. God knew the springs were there all the time.[4]

Some, however, have tried to show that statements made in the Bible are incorrect. For example, some have said that the Bible teaches that the earth is flat. The scripture that they take out of context is Isaiah 11:12 which states that God will "gather together the dispersed of Judah from the four corners of the earth." A similar expression would be "from the east, west, north, and south." Neither expression means the earth is flat. Neither does the expression prevent those from the southeast, southwest, northeast, northwest, north central or any other part of the earth from being gathered. The expression "from the four corners of the earth" means from all over the earth. I don't know of any of the above critics who mention that the same author in Isaiah 40:22 states, "It is He who sits above the circle of the earth." Photographs from space have verified that the perimeter of the earth is circular.

Are some subjects labeled as science really "science falsely so called"?

Much of the content of so called science courses such as geology and biology is falsely called science. Many of the opinions presented regarding the origin of the earth or life are presented as scientific knowledge but these ideas cannot be observed or verified nor are they subject to experimentation. Thus these opinions are not scientific but rather "science falsely so called." For example, the *Science Framework for California Public Schools Kindergarten Through Grade Twelve* properly states "Science aims to be testable, objective and consistent. If an idea cannot (even potentially) be so tested then it is outside the realm of science. Explanations of nature must be based on natural phenomena and observations, not on opinions or subjective experiences. One good control of scientific objectivity is the repeatability of science; that is any observation ought to be repeatable and capable of being confirmed or rejected by other scientists. A scientific explanation must clearly agree with all the observable facts better than alternative explanations do. Because science is open-ended, those who would practice or understand it must be open-minded." However, two sentences later the *Framework* states that the age of the earth is 4.54 billion years old. Is such an idea observable, repeatable, or capable of being confirmed? The next sentence says that as a new age of the earth is determined "we know that the new value will not be 10,000 years or 100 billion years."[5] Is that being open-minded to say we know something ahead of time? What it really says is we are closed minded to any idea, which promotes the idea of a young earth. Later, on page 17, a statement is made that "science never commits itself 'irrevocably' to any fact or theory, no matter how firmly it appears to be established in the light of what is known." How contradictory can they get? Also the *Framework* states that the age of the earth "has changed by only 0.01 billion years in over three decades."[6] It has been noted, however, that "over the past century, the estimated age of the earth has doubled approximately every twenty years."[7]

The California *Science Framework* appears to be an apologetic for Darwinian evolution. In fact the *Framework* is little more than a primer on evolution. It uses the term "evolution" 47 times.[8] The following quotes serve as examples: "Evolution is the central organizing theory of biology and has fundamental importance in other sciences as well. It is an accepted scientific explanation and therefore no more controversial in scientific circles than the theories of gravitation and electron flow. Evolution is not confined to the earth and its systems but extends to the entire universe. Evolution embodies history and therefore is a part of every discipline in which history has a role. In order to teach life science, earth science, or astronomy, evolution should be a fundamental, central concept of the curriculum."

Are science textbooks loaded with non-scientific ideas?

California has the largest population of all the states. Books sold to public schools in California must fit the guidelines of the *Framework*. Since publishers don't want to incur the expense of printing many editions, they developed books to meet the demands of California, which is their major market. Unfortunately these books with evolution as one of the major themes are then sold throughout the country.

I was asked to evaluate the book *The Nature of Science.*[9] On page 11 when introducing the subject of "What is Science" it states that Dr. W. P. Coombs, Jr., took "one look" at some scratches in rocks and "concluded that the scratches were made by an animal having three toes with sharp claws. They were clearly the work of the meat-eating dinosaur called Megalosaurus." He also claimed it was a "swimming, meat-eating dinosaur." Were his conclusions, such as meat-eating and swimming, observable? Was this repeatable, verifiable or subject to experimentation? Is this science? At best it is science fiction. Suppose you poured a concrete driveway and the next day saw some footprints in the driveway. Could you with "one look" scientifically conclude that meat-eating, swimming Southeast Asians made these footprints? I'm afraid you would quickly be labeled as a racist bigot. Can you, by footprints, determine if a person or animal eats meat? I can't even tell if a living person is meat-eating by looking at his teeth unless meat is found between the teeth. The person might be a vegetarian. Can you tell by footprints whether something can swim or not? This book says Dr. Coombs made "an important scientific discovery." This is not a scientific discovery but rather "one person's imagination." On page 27 the book states "Believe it or not, many scientists search for truth of nature without ever performing experiments." The book goes on to say "much of what we know about evolution is based on Darwin's work. Yet Darwin did not perform a single experiment!" The entire question of origins is not scientific because it is not "observable or testable." I also question why the book chooses a man who performed no experiments as an example of a scientist. Why didn't they choose a man like Isaac Newton who is considered by many as the greatest scientist that ever lived? His contributions include the law of universal gravitation, the three laws of motion, and studies in light and color. Space does not permit me to mention many other non-scientific phrases found in *The Nature of Science.*

It has been said that if you believe that a frog turned into a prince by a kiss, you believe in a fairy tale. However, if you believe that a frog turned into a prince over millions of years, you believe in evolution. Certainly the theory of evolution is falsely called science. Perhaps it could be called "Science Fiction."

Why is it important to believe the Genesis record of creation?

Jesus said, "If you do not believe his [Moses] writings, how will you believe My words?" (John 5:47). Also if a person doesn't believe the words of Moses, he will not be persuaded even if one rose from the dead (Luke 16:31). If the first Adam was not real and if the fall of man into sin did not take place, then neither is the second Adam (Jesus Christ) real and there is no need of a Saviour. The book of Genesis is not only foundational for the creation account but also for most other major doctrines regarding sin and salvation. If you can't believe in the God of creation, how will you believe in the incarnation or the God of salvation?

---

[1] *Webster's New World Dictionary of the American Language*, Second College Edition, The World Publishing Co., New York and Cleveland, 1968.

[2] Duane T. Gish, "Modern Scientific Discoveries Verify the Scriptures," Impact Article No. 219, Institute for Creation Research, El Cajon, CA, September 1991.

[3] Duane T. Gish, "Remarkable Scientific Accuracy of Scripture," Institute for Creation Research, El Cajon, CA, Science, Scripture and Salvation, Weekly Broadcast No. 165 aired weekend of July 8, 1989.

[4] Steven A. Austin, "Springs of the Ocean," Impact Article No. 98, Institute for Creation Research, El Cajon, CA, August 1981.

[5] Science Framework for California Public Schools Kindergarten through Grade Twelve, Bureau of Publications, California Department of Education, P.O. Box 271, Sacramento, CA, 1990, 14-15.

[6] Ibid., 17

[7] Wayne Jackson, *Creation, Evolution, and the Age of the Earth*, Courier Publications, P.O. Box 55265, Stockton, CA, 1989, 2.

[8] Bruce Schweigerdt, Personal correspondence

[9] *The Nature of Science*, Prentice-Hall Inc., Englewood Cliffs, NJ, 1993,11-27.

2 / Can a Christian Be a Scientist?

A fact not mentioned in most textbooks is that many of the greatest founders of modern science were Bible-believing Christians. (What other kind of Christians are there?) Let's look at just a few of these.

Johannes Kepler (1571-1630) is considered to be the founder of physical astronomy.[1] He discovered the laws of planetary motion, which are still used in calculating orbits of satellites and in mapping routes for space travel. His first serious scientific treatise, *The Mystery of the Universe*, closed with a magnificent hymn of praise to the Creator. Kepler was a devoted family man who sought to give his children a Christian upbringing. To aid their understanding he wrote Bible Study Guides. One of these guides, "The Body and Blood of Jesus Christ Our Savior" is still preserved in the University of Tubingen library. A crater on the moon was named after him as well as some monuments and museums. Kepler stated, "Let my name perish if only the name of God the Father is thereby elevated." Kepler is also called "the father of modern optics" because of his mathematical analysis of lenses and mirrors.[2]

Robert Boyle (1627-1691) is generally credited with being the father of modern chemistry who guided the great transition from alchemy to true chemistry. Boyle was the first to define elements and compounds in terms of experimental observations. He is best known for his study of gases under pressure, but he also investigated the properties of a vacuum, combustion, propagation of sound and the effects of reduced air pressure upon breathing.

Throughout his life Boyle read the Bible each morning. He wrote many tracts and numerous essays showing that science and religion both were a study of God's creation. During his later years, he supported missionary endeavors in many countries. Also he commissioned translations of the four Gospels and the Book of Acts into Turkish, Arabic and Malayan. Because of his burden for his fellow Irishmen, he financed a new Irish translation of the entire Bible. Thousands of these Bibles were distributed at Boyle's expense. In his will he provided funds for the "Boyle Lectures," a series of eight sermons, to be delivered each year to demonstrate that Christianity is intellectually defensible and far more reasonable than the various philosophies that sought to discredit it.[3]

Isaac Newton (1642-1727) is judged to be the greatest scientist who ever lived. His book *Principia* (The Mathematical Principles of Natural Philosophy) is generally regarded as the greatest book on science of all time. In this book are stated the law of universal gravitation and the three laws of motion. He calculated the masses of the sun and planets in terms of the earth's mass. Another book he wrote, *Opticks*, (which summarized his studies of light and color) was as much of a breakthrough in optics as *Principia* had been in dynamics. He also made fundamental contributions to pure mathematics,

9

discovering the binomial theorem, and made the first reflecting telescope. For his accomplishments, he became the first person to receive knighthood from the Queen of England for scientific achievement.[4]

What is lesser known is that science was not Isaac Newton's chief interest. He spent more time studying the Bible than he did investigating nature. He wrote numerous Gospel tracts and two books on religious subjects: *The Prophecies of Daniel* and *Chronology of Ancient Kingdoms*. He is quoted as saying, "Atheism is so senseless. When I look at the solar system, I see the earth at the right distance from the sun to receive the proper amounts of heat and light. This did not happen by chance. The motions of the planets require a Divine Arm to impress them."[5]

Science historians regard Michael Faraday (1791-1867) as the greatest of the experimental physicists. The transformer, the electric motor and the electric generator are some of his inventions. He discovered the laws of electrolysis and two basic units in physics are named in his honor. The "faraday" is a unit of electrical quantity and the "farad" is a unit of electrical capacity. Terms such as "magnetic field" and "lines of forces" were products of his thinking. He also contributed to our knowledge of polarized light and the liquefaction of gases. His discovery of benzene laid the foundation for aromatic organic chemistry.

Faraday was a devoted student of the scriptures and wrote in the margins of his Bible nearly three thousand notations in the form of study aids, comments and cross-references. A reporter once asked Faraday about his speculations on the hereafter. He replied that he had no speculations but rather he rested on certainties and quoted II Timothy 1:12, "I know whom I have believed, and am persuaded that he is able to keep that which I have committed unto him against that day."[6]

In the references listed below one can read about many other Bible-believing scientists such as Louis Pasteur (pasteurization, rabies vaccination), James Maxwell (electricity and the kinetic theory of gases), Lord Kelvin (absolute temperature), Gregor Mendel (heredity), William Ramsay (noble gases — helium, neon, argon, krypton and xenon), Carolus Linnaeus (classification system of plants and animals), Joseph Henry (electromagnetism, unit of electrical inductance), Francis Bacon (scientific method) and Samuel Morse (telegraph). The researches and analyses of the others led to the very laws and concepts of science, which brought about our modern scientific age. Many of these men were strong opponents of Darwinism.[7]

Not only were the great scientists of the past believers in the Bible, but there are thousands of scientists today who believe in the Bible. Julian Huxley was lying when he said, "No serious scientist would deny the fact that evolution has occurred..."[8] It is absolutely not true that a scientist cannot be a Christian or a Christian cannot be a scientist. The Creation Research Society, Institute for Creation Research and other similar organizations are composed of many scientists who embrace the inspiration of the Bible.

---

[1] Henry M. Morris, *Men of Science - Men of God*, Master Books, A Division of Creation Life Publishers Inc., El Cajon, CA, 1982, 11.

[2] Emmett L. Williams, and George Mulflinger, *Physical Science for Christian Schools*, Bob Jones University Press, Greenville, S.C., 1974, 363-366.

[3] Ibid., 127-128.

[4] Basic Science Physics Pace No. 135, Reform Publications, Inc., 1987, 2-5.

[5] John Hudson Tiner, College Physical Science Pace 50003, Accelerated Christian Education, Inc., 2600 ACE Lane, Lewisville, TX, 1980, 27-28.

[6] Williams and Mulflinger, 421-422.

[7] Henry M. Morris, "Bible Believing Scientists of the Past," Impact Article No. 103, Institute for Creation Research, El Cajon, CA, January 1982.

[8] Henry M. Morris, *The Twilight of Evolution*, Baker Book House, Grand Rapids, MI, 1963, 25.

3 / Benefits of Science

What are the benefits of science to mankind?

Because of much "science falsely so called" there is a tendency among some Christians to "bash" all science and scientists. This should not be. All one has to do is look around and see the benefits of science and technology. The following is a partial list of some of these benefits:

Electricity: lights, washers, dryers, microwaves, dishwashers, disposals, garage door openers, motion detection lights, automatic sprinklers, elevators, escalators, refrigerators, air conditioners and overhead projectors.

Transportation: Automobiles, trains and airplanes.

Communications: Telegraph, telephones, cellular phones and the Internet.

Electronics: Videos, compact discs, fax machines, computers, tape recorders, radios, laser printers, scanners and remote controllers.

Plastics: fabrics, shoe soles, PVC pipes, Teflon pans, polyethylene Tupperware, see through packaging, nylon, latex paints and carpeting.

Agriculture: fertilizers, insecticides and seedless fruits.

Beauty Products: hair spray, shampoo and deodorants.

Medicine: Blood thinners, blood pressure reducers, pain reducers, vaccines, x-rays and chemotherapy. The average lifespan has risen from about thirty in 1750 to over seventy now. Some of this is a result of medical science.

4 / Original Creation

Who created all things?

To create means to make out of nothing. Genesis 1:1 declares, "In the beginning God created the heavens and the earth." God was before anything material existed. God "made out of nothing" the universe. "For by him were all things created that are in heaven, and that are in earth, visible and invisible, whether they be thrones, or dominions, or principalities, or powers: All things were created by him, and for him" (Colossians 1:16). There is no disagreement among the various authors of the Bible as to who was the creator as shown by the following scriptures:

1. *David.* "The heavens declare the glory of God; and the firmament shows His handiwork" (Psalm 19:1).

2. *Solomon.* "Remember now your Creator in the days of your youth..." (Ecclesiastes 12:1).

3. *Isaiah.* "Thus says God the LORD, Who created the heavens and stretched them out, Who spread forth the earth and that which comes from it, Who gives breath to the people on it, and spirit to those who walk on it" (Isaiah 42:5).

4. *Jeremiah.* "Ah, Lord GOD! Behold, You have made the heavens and the earth by Your great power..." (Jeremiah 32:17).

5. *Amos.* "For behold, He who forms mountains, and creates the wind, Who declares to man what his thought is...The LORD God of hosts is His name" (Amos 4:13).

6. *Nehemiah.* "You alone are the LORD; You have made heaven, the heaven of heavens, with all their host, the earth and everything on it, the seas and all that is in them, and You preserve them all. The host of heaven worships You" (Nehemiah 9:6).

7. *Malachi.* "Have we not all one Father? Has not one God created us?" (Malachi 2:10).

8. *Job.* "Where were you when I laid the foundations of the earth? Tell Me, if you have understanding" (Job 38:4). In chapter 38 and 39, God asks Job a number of tough questions showing Job that man's knowledge of creation is so limited.

9. *Mark.* "...such as has not been since the beginning of the creation which God created until this time ..." (Mark 13:19).

10. *Luke.* "Lord, You are God, who made heaven and earth and the sea, and all that is in them" (Acts 4:24).

11. *John.* "You are worthy, O Lord, to receive glory and honor and power; for You created all things, and by Your will they exist and were created" (Revelation 4:11).

The only evolutionary statement in the Bible that I know of is when Moses questioned Aaron about the golden calf and Aaron replied, "I cast it (gold) into the fire, and this calf came out" (Exodus 32:24). We know he was lying because Exodus 32:4 records that he received the gold from the people, he "fashioned it with an engraving tool, and made a molded calf." Evolutionists who claim that the creation came about by time and chance are also propagating a lie and are falsely called scientists.

How did the writers of the Bible get their information?

"All Scripture is given by inspiration of God" (II Timothy 3:16). The writers of the Bible were holy men of God who "spoke as they were moved by the Holy Spirit" (II Peter 1:21). Since God was the only one there at the time of creation and it was His idea and plan, He is the one source for this information. We believe in the prophecy of the Bible where God reveals the future. It is just as easy for God to reveal the past as it is the future.

What was the order of His creation?

Genesis 1:1 declares, "In the beginning (time) God created the heavens (space) and earth (matter/energy)". True science deals with the present (that which is observable). Thus we must go to history to find out about creation that occurred in the past. Those who were there and observed the happenings write the most accurate records of history. The Spirit of God the creator moved upon the writers of the Bible to record the creation events. Genesis 1:2 seems to indicate that the earth originally was in darkness and was covered by water.

On the first day God said, "Let there be light." What was that light? We know that it was not the sun because the sun was not created until the fourth day. Whereas we do not know for sure what that light was, we do know there are many other sources for light other than the sun. For example, electricity can produce light. There are also other forms of radiation that give off light. We also know that God is light (I John 1:5) and that light could emanate from Himself such as it did on the mountain of transfiguration. With this light God divided the light from darkness and called the light Day and the darkness Night. And the evening and the morning were the first day. It seems that during this day God gave the earth a spin and it has been spinning ever since to give us day and night. A "Big Bang Explosion" couldn't have formed the spinning motion of the earth because a radial explosion could not cause a spinning or orbiting motion in spherical particles. Photos from space show that the perimeter of the earth is a circle.

14

On the second day God made a firmament which divided the waters which were under the firmament from the waters which were above the firmament. The waters, which were above the firmament, have been called by many a "vapor canopy." You can read more about this in chapter five and six.

On the third day God gathered the waters together (seas) and caused the dry land (earth) to appear. For further details on this see chapter eight. On this day He also made the great variety of plant life, fully developed, yielding herbs and fruit. It would appear that these fruit trees were many years old in that they were already bearing fruit, but God made them in one day.

On the fourth day God made the sun, moon and stars to divide the day from the night. Apparently God replaced the light of Day One with the sun and moon. He also made them for signs and for seasons. It was apparently at this time that God put the earth in orbit around the sun and the moon around the earth. Psalm 19:1-4 lets us know that the sun, moon and stars show God's handiwork and are to give witness to all the earth of the creator. What keeps the earth spinning and revolving around the sun rather than stopping or flying out into space? It is "by Him all things consist" (Colossians 1:17).

On the fifth day God made the water dwelling animals, such as the whales, and the birds which fly.

On the sixth day God made the earth-dwelling animals, such as the cattle and creeping things, which could refer to insects and snakes. Last of all on the sixth day God made man, the ultimate of his creation, after His own image. I believe this first man was a fully developed man on the day of his creation. Perhaps he looked thirty years old. Creation of fully developed things results in a concept of apparent age. Thus the trees, stars, man and all of creation might appear to be old to us because of our realm of thinking in which we are used to seeing things grow over a period of time. Even if the stars are millions of light years away, God had to create the light beams as well as the stars in order for them to help rule by night (Psalm 136:9). The light was created with the stars and instantly spread across space. When God speaks, "Let there be," it can happen instantaneously. The stars must have been visible from the moment of their creation on the fourth day. In our day we do not see new stars "turning on" in the night sky as if their light has finally reached the earth.[1] Creationists have no problem with the fact that stars differ in size, temperature and brightness. "One star differs from another star in glory" (I Corinthians 15:41).

How can the original creation be described?

When God created the light He saw, "...that it was good" (Genesis 1:4). When God created the dry land and seas He "...saw that it was good" (Genesis 1:10). When God created the grass, herbs and fruit trees, "God saw that it was good" (Genesis 1:12). When God created the sun, moon and stars "God saw that it was good" (Genesis 1:18). When God created the swimming creatures and fowl He also "...saw that it was good" (Genesis 1:21). And when God created the beasts, cattle and creeping things "God saw that it was good" (Genesis 1:25). After the creation of man when God saw everything that He had made He declared the complete creation was "...very good" (Genesis 1:31). The only thing that was stated as "not good" was that Adam was alone. God took care of this by creating a helpmeet for him (Genesis 2:18). Since Lucifer and the angels were part of the invisible creation it seems that Lucifer had not fallen at this time or else God could not have called all of his creation very good. God didn't say everything He made was good except for Satan and his rebellious angels. Satan must have fallen shortly thereafter sometime prior to the third chapter of Genesis.

The original creation has been called a perfect paradise. There were plenty of herbs and fruit trees and Adam's job was to tend and keep the garden of Eden which must have been quite easy (Genesis 2:15). It must have been an ideal climate since Adam and Eve were naked.

Who created man?

If God created all things this includes the creation of man. Genesis 1:27 declares that, "God created man in His own image."

How did God make man?

Genesis 2:7 declares that, "the Lord God formed man of the dust of the ground, and breathed into his nostrils the breath of life..." Every atom in the human body, such as carbon, hydrogen, oxygen, calcium and nitrogen, can be found in the earth which contains minerals, elements and vegetation.

Who were the first man and woman?

I Corinthians 15:45 plainly declares that the first man was Adam. Also Genesis 3:20 lets us know that Eve was the mother of all living.

What did the first man look like?

Genesis 1:27 declares that the first man was made in the image of God. He did not look like some transitional form between an ape and man, but rather he looked like Jesus Christ who was the image of the invisible God (Hebrews 1:3, Colossians 1:13-15, II Corinthians 4:4). Romans 5:14 lets us know that

16

Adam was "the figure of him that was to come" (KJV). Thus Adam looked like Jesus Christ who was yet to come. Adam did not look like a "prehistoric man," as evolutionary artists would draw, but rather as Jesus Christ, fully developed and fully intelligent. Adam must have been intelligent to name all the animals.

When was man created?

Archbishop Ussher[2] determined the approximate date of the creation of man by a combination of genealogies and history. Genesis 5:1-32 gives the pre-flood genealogies from Adam to Noah. For example, Adam was 130 years old when he begat Seth, and Seth was 105 years old when he begat Enos. Thus by addition one can determine that Enos was born 235 years after Adam was created. By totaling all these years it can be calculated that Noah begat Shem, Ham and Japheth 1556 years after the creation of Adam.

NAME OF PERSON OR EVENT	YEAR OF BIRTH FROM CREATION	SCRIPTURAL REFERENCE IN GENESIS
Adam	0	1:26, 27, 2:7, 5:1
Seth	130	5:3
Enos	235	5:6
Cainan	325	5:9
Mahalaleel	395	5:12
Jared	460	5:15
Enoch	622	5:18
Methuselah	687	5:21
Lamech	874	5:25
Noah	1056	5:28, 29
Shem, Ham, Japheth	1556	5:32
"Flood"	1656	7:6

By using the post-flood genealogies in Genesis 11:10-26 one can determine that Abraham was born 1948 years after Adam. Also Genesis 21:5 and 25:26 establish the date of Jacob's birth as 2108 years after Adam and Genesis 48:9 shows that Jacob went down to Egypt 2238 years after Adam. Exodus 12:40 indicates that the children of Israel were in Egypt 430 years, which would establish the exodus at 2668 years after Adam. These genealogies, as well as Genesis 29:35; 38:29; Ruth 4:18-22; Matthew 1:1-16; Luke 3:23-38, bring us up to the time when history and archaeology can fairly accurately determine dates.

NAME OF PERSON OR EVENT	YEAR OF BIRTH FROM CREATION	SCRIPTURAL REFERENCE IN GENESIS
Arphaxad	1658	11:10
Salah	1693	11:12
Eber	1723	11:14
Peleg	1757	11:16
Reu	1787	11:18
Serug	1819	11:20
Nahor	1849	11:22
Terah	1878	11:24
Abram (Abraham)	1948	11:26
Isaac	2048	21:5
Jacob	2108	25:26
Jacob Went To Egypt	2238	47:9
Moses Left Egypt	2668	Exodus 12:40

By this combination of history, archaeology and genealogy one can establish that God created Adam approximately 4000 years B.C. and that the Flood occurred about 2500 years B.C. As a confirmation to the importance of these genealogies, Luke traces the genealogies of Jesus Christ clear back to Adam, while Matthew, who was writing primarily to the Jews, gives the genealogy from Abraham to Jesus. In the evolutionary scheme of things it is contended that the earth is 4.5 billion years old but man is 3.6 million years old only $1/1250^{th}$ of the age of the earth. Thus according to evolutionists, on a scale of 1000 periods, man evolved in the last period. Jesus, however, said "But from the beginning of creation God made them male and female" (Mark 10:6). Thus Jesus put the creation of man in the same time frame as the creation of the earth.

Some Biblical scholars use types to confirm the date of creation. The Passover lamb was kept four days (from the tenth to the fourteenth day) before it was slain. The Passover lamb was clearly a type of Jesus Christ. II Peter 3:8 says "with the Lord one day is as a thousand years." By using this comparison, then four days would be 4000 years or, in other words, Jesus Christ, the second Adam, was kept approximately 4000 years before being slain. Other scriptures used in this comparison are Hosea 6:2, Luke 11:35, Matthew 20:2 and Luke 13:32.

Other Biblical scholars point out that abridgement of genealogies is common in scripture and that the term "son" at times means descendant. For example, Christ is called the son of David. Some of the genealogies have names omitted or added when compared to other genealogical lists. Some believe differences occurred when translating between the Hebrew and Septuagint text. Others feel that the purpose of the genealogical tables were not to give an exact date but to give us the names of the important men that lived during these periods. These scholars acknowledge that God gave the ages of these men to show us that the earth is not billions of years old.[3] Many fundamental creationists will allow the age of man to be up to 10,000 years rather than 6,000 years.[4] Acknowledging this potential difference does not alter any of the principles set forth in this book.

[1] Donald B. DeYoung, *Astronomy and the Bible*, Questions and Answers, Baker Book House, Grand Rapids, MI, 1990, 80-81.

[2] Whitcomb, John C., Jr., and Morris, Henry M.; *The Genesis Flood - The Biblical Record and Its Scientific Implications*, Baker Book House, Grand Rapids, MI, 1961.

[3] John W. Klotz, *Genes, Genesis and Evolution*, Concordia Publishing House, St. Louis, MO, 1955, 92-96.

[4] Harold S. Slusher in William J. J. Glashouwer and Paul S. Taylor, writers, The Earth, A Young Planet?, Mesa, AZ: Eden Films and Standard Media, 1983 (Creationist Motion Picture).

5 / World Before the Flood

What three states of the earth are described in II Peter chapter 3?

1. In verse 6 Peter stated that "the world that then was, being overflowed with water, perished" (KJV) This is referring to the world *before the Flood*. The world we now live in is not the same as it was before the Flood. There is no telling what happened when "all the fountains of the great deep were broken up, and the windows of heaven were opened" (Genesis 7:11). The great damage done by geysers, volcanoes, earthquakes and resulting tsunamis must have been immense. Great numbers of plants and animals were buried as evidenced by huge fossil graveyards, coal seams and oil fields. Also immense lava beds, such as the Columbia Plateau, and huge thrust faults, such as the Matterhorn in the Alps, give evidence to a great catastrophe.

2. In verse 7 he stated "But the heavens and the earth which are now...are reserved unto fire against the day of judgment..." (KJV). This refers to the earth that we *are now* living on.

3. In verse 13 it states that we "Look for new heavens and a new earth wherein dwelleth righteousness" (KJV). This refers to the earth that *will be* formed after the present earth is destroyed by fire.

What additional things were different about the earth before the Flood?

1. Originally man and all the animals were vegetarians (Genesis 1:29, 30). Since death entered the world by sin and man had not sinned as yet there would have been no death prior to Adam's sin (Romans 5:12). But what about plants or microorganisms that would have died in the Garden of Eden? Plants don't have consciousness, blood, working muscle tissue or breath. Plant life, which was created on the third day, is different from animal life, which was created on the fifth and sixth day. It was the death of animal type life that began after Adam's sin.[1] It appears that man did not eat animals until after the Flood when God said "Every moving thing that lives shall be food for you. I have given you all things, even as the green herbs" (Genesis 9:3).

2. The world before the Flood seems not to have experienced rain. Genesis 2:5 states "...for the Lord God had not caused it to rain on the earth..." A mist from the earth watered the face of the ground. Without rain, erosion was non-existent and there probably were no droughts, deserts or wasteland. Hebrews 11:7 states that "Noah was warned of God of things not yet seen." This was probably referring to the rain, which the people of

Noah's day had not seen. The rainbow, which occurred after the Flood (Genesis 9:13-14), was a new phenomenon to Noah and his family.

3. Extreme seasons apparently did not exist until after the Flood. Although there were some kind of seasons after the creation of the sun (Genesis 1:14) cold, heat, summer and winter were not mentioned previous to Genesis 8:22.

4. There appears to have been a warm, uniform climate. Adam and Eve were both naked until they sinned (Genesis 2:25). They did not seem to need clothing because of cold or heat, but after they sinned, clothing became a requirement for modesty. Fruit was in abundance and the fossil record indicates tropical plants and animals were all over the world, including Greenland and Anarctica.[2] Palm trees grew all over the United States. The state of New York was as warm as Florida. Fossil fuels are found in such unlikely spots as Alaska's North Slope.[3] Many people feel that the warm climatic conditions prior to the Flood were caused by the dividing of the waters from the waters as recorded in Genesis 1:6. When God created the earth water covered the earth. On the second day God divided the waters from the waters and put a firmament between. The picture would be that of a spherical earth covered with water, surrounded by another spherical water layer with space between the earth and the outside spherical layer. This outside spherical layer has been termed a vapor canopy by many creationists. The canopy probably produced a "greenhouse effect," warming the atmosphere just as the glass of a greenhouse tends to raise the temperature of the interior. This water canopy would produce a globally, almost uniformly, warm climate which can account for tropical plants all over the earth, even in the regions which are now arctic regions since the precipitation of the canopy. This vapor canopy also could account for the temperate climate required for ferns, lush vegetation and dinosaurs around the world. Also, it could have prevented winds, storms and seasonal variation. The continual growing season probably resulted in larger specimens of plants and animals. This canopy probably provided a warm, pleasant, healthful environment.

5. Dinosaurs roamed the earth. The bones of dinosaurs have been found in all the continents except Anarctica. It is felt that some weighed as much as 100 tons. An adult bull elephant today can consume 300-600 pounds of fodder daily.[4] A dinosaur weighing over fifteen times as  much would probably have had to consume fifteen times that amount of food. Lush vegetation must have covered the earth before the Flood in order for these creatures to survive. It is possible also that the increased air

pressure due to the vapor canopy was what enabled flying reptiles to fly.[5] It is doubtful these creatures could have flown in today's atmosphere.

6. Men lived long ages. Methuselah, for example, lived 969 years (Genesis 5:27). The vapor canopy probably shielded humans from harmful somatic (non-hereditary) and genetic (hereditary) effects of intense radiations. It has been shown, from a study of more than 82,000 physicians, that the life span of those who had exposure to radiation was over five years less than those without exposure to radiation.[6] The shielding effect of the canopy enabled antediluvian persons to live to great ages as recorded in Genesis 5.

What health benefits might have resulted from an increase in pressure due to a vapor canopy?

One of the objections to the vapor canopy theory is that the additional water vapor above the troposphere would increase the barometric pressure to levels that would be lethal to human life. However, this would depend on how much water was in the canopy. It has been shown, instead, that some increase in pressure would be beneficial. Dr. Edgar End, at the University of Wisconsin, has demonstrated that inhaling hyperbaric oxygen, administered in a pressure chamber, will restore memory, increase energy and add zest to older men and women. It has also been shown that hyperbaric oxygenation frequently reverses senility, helps stroke victims, improves eye sight, reduces the time required for healing severe burns and saves victims poisoned by carbon monoxide.[7] A severe cut on the hand of an aquanaut was healed in twenty-four hours while submerged in a diving bell at a pressure of ten atmosphere.[8]

---

[1] Jim Stambaugh, Science, Scripture and Salvation, Tape No. 476, Institute for Creation Research, El Cajon, CA, Saturday June 24, 1995.

[2] Ricki D. Pavlu, *Evolution: When Fact Became Fiction*, Word Aflame Press, Hazelwood, MO, 1986, 28.

[3] Donald B. DeYoung, *Weather and the Bible, 100 Questions and Answers*, Baker Book House, Grand Rapids, MI, 1993, 110.

[4] Pavlu, 36.

[5] Larry Vardiman, "The Sky Has Fallen," Impact Article No. 128, Institute for Creation Research, El Cajon, CA , February, 1984.

[6] John C. Whitcomb, Jr., and Henry M. Morris, *The Genesis Flood - The Biblical Record and Its Scientific Implications*, Baker Book House, Grand Rapids, MI, 1961, 400.

[7] Henry M. Morris, *The Biblical Basis for Modern Science*, Baker Book House, Grand Rapids, MI, 1984, 278-280.

[8] Vardiman, Impact Article No. 128.

6 / The Great Flood

Other than creation itself, what was the most significant event that affected the geologic formations of the earth?

The great flood, often called Noah's Flood, is described in Genesis 7 and 8. The Flood must have restructured the entire face of the earth. The crust of the earth seems to have been distorted, fractured, elevated, depressed and contorted in almost every conceivable way.

Where did all the waters come from?

After Noah entered the ark, "all the fountains of the great deep were broken up, and the windows of heaven were opened" (Genesis 7:11). The "fountains of the deep" were perhaps subterranean geysers, springs, volcanoes and earthquakes under the water which caused tsunamis, while the "windows of heaven" was probably the vapor canopy (waters which were above the firmament) collapsing. This vapor canopy could have accounted for the large amount of water required for a forty day rain recorded in Genesis 7:11, 12. Authorities have estimated that if all the water vapor in the atmosphere would fall as rain it would amount to a 2-7 inch sheath of water.[1,2] Thus a completely different mechanism than rain from the atmosphere must have been responsible for a global rain of forty days and an earth-covering flood. Perhaps also the ocean basins rose and the mountains lowered.

What indicates that the whole earth was covered by the Flood?

The Bible plainly teaches that it was a worldwide flood, in that all the high hills under the whole heaven were covered and that "The waters prevailed fifteen cubits upward, and the mountains were covered" (Genesis 7:19-20). Thus the tops of the highest mountains were submerged under water by at least twenty-two feet. Water had to be deep enough so the bottom of the ark would not hit the tops of the mountains. If the Flood had been local, there would have been no need for an ark. In 120 years Noah and the animals could have walked to a higher elevation. Also, the primary goal was to destroy sinful mankind (Genesis 6:7), which would not have occurred with a local flood. The scripture declares that every living substance which was on the face of the earth was destroyed, except for Noah and those with him in the ark (Genesis 7:23).

What else shows that the Flood was world-wide rather than local?

1. There are at least thirty expressions of universality indicating a world-wide flood, such as "All flesh died" (Genesis 7:21) and "All in whose nostrils

was the breath of the spirit of life, all that was on the dry land, died" (Genesis 7:22).

2. The worldwide character of the Flood was proclaimed in later parts of Scripture.

3. Peter declared only eight souls were saved by water (I Peter 3:20).

4. And "the world that then existed perished, being flooded with water" (II Peter 3:6).

5. If the Flood was local, the ark was ridiculously large for its purpose.

6. Three times God promised never again to smite every living thing by a flood. (Genesis 8:21, 9:11, 15). Why would God promise never again to do something if He never did it in the first place?[3]

What evidence is there today of such a great flood?

1. Nearly all of the great mountain areas of the world have been found to have marine fossils near their summits. There is no conclusion possible other than that the mountains were once under water and have all been uplifted, essentially simultaneously, which accords well with the biblical Flood[4,5]

2. Fossilization requires a sudden burial. Coal seams, oil fields and natural gas deposits are specific examples of fossil fuels.

3. Most of the rocks on the earth's surface are sedimentary rocks that require heavy rain for erosion, transportation and deposition of sediment.

4. Anthropologists have collected over two hundred myths and legends of the Flood from civilizations all over the world. Since all humans are

descendants of Noah, one would expect to find some stories of the Flood in ancient records. Such is the case. These flood legends consistently show that a worldwide flood destroyed both man and animals, a vessel of safety was provided and only a few people survived the Flood.[6,7]

[1] Whitcomb and Morris, *The Genesis Flood*, 121.

[2] Ibid., 77.

[3] John D. Morris, "Scripture and The Flood," Notes presented at Institute of Creation Research Summer Institute, Northwestern College, St., Paul, MN, July 11, 1988.

[4] Pavlu, 57.

[5] Morris, *Biblical Basis*, 259.

[6] Paul S. Taylor, *The Illustrated Origins Answer Book*, Eden Productions, P.O. Box 41644, Mesa, AZ, 1990, 112-113.

[7] John Rajca, "Flood Accounts Around the World," Science, Scripture and Salvation Broadcast No. 170, Institute for Creation Research, El Cajon, CA, August 12, 1989.

7 / Capacity and Capabilities of Noah's Ark

Was the ark large enough to hold all the animals?

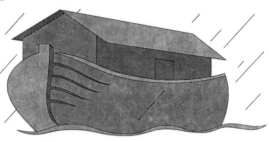

The ark is the only ship in the Bible where the dimensions are given. The length was to be 300 cubits, the breadth 50 cubits and the height 30 cubits (Genesis 6:15). Assuming the cubit to be one and a half feet, we can calculate the volume, 450 x 75 x 45 = 1,518,750 ft³, which is equivalent to 569 railroad stock (box) cars.[1] It had one window, one door and three stories. The deck area was more than 100,000 ft², which is greater than twenty basketball courts.[2]

Noah was to bring two of every sort (kind) into the ark and of every clean beast he took seven so he would have some for sacrifices. He brought fowls, beasts (cattle) and creeping things but did not have to bring fish or other things that live in the sea. He was to bring food for the animals. Note that all he had to bring were two of a "kind," such as two dogs. He didn't have to bring great Danes, cocker spaniels, bulldogs, etc. Eliminating aquatic species, such as fish, lobsters and turtles, it is estimated that there were 17,500 kinds on the earth with approximately 50,000 animals on board.[3] If the average size of the animals was no greater than a sheep and a boxcar would hold 240 sheep, then 208 boxcars would be required. This would be about 36.5% of the ark's capacity. Thus the ark was certainly large enough for the animals, the food and Noah's family.

Was the ark capable of withstanding the violent storms during the Flood?

The ark was essentially a huge box. It was designed for floating and cargo capacity, not sailing or speed. It was like a huge barge, not a sailboat. It was made of strong gopher wood and pitched within and without. The pitch was some kind of waterproofing material of unknown origin. It was designed to float and not capsize under the impact of great waves and winds. Mathematical and buoyancy calculations show that until the ark would have been turned vertically, the buoyant force would produce a "righting couple" that would act to restore the vessel to its upright position.[4] Commander Lee Touchberry of the U.S. Navy took the ark's dimensions and put them into the Navy's computerized programs. Calculations show that the dimensions of the ark were a perfect match with the Navy's ideal ship dimension ratio.[5]

The Scripps Institute of Oceanography at La Jolla, California, tested a scale model of the ark in a large wave tank during the filming of the movie "In Search of Noah's Ark." By mechanical means, large waves were produced in the tank to test the capability of the ark in simulated waves greater than any ever experienced in the ocean. It was found that the ark was impossible to capsize.[6] The divinely given dimensions were ideal for its purpose!

How did Noah get animals into the ark?

God directed the animals to the ark. "...two of every kind will come to you..." (Genesis 6:20). With a warm, uniform climate prior to the Flood, all of the kinds of animals probably lived near where Noah was building the ark. Also, authorities have noted that animals have an innate sense of imminent danger. God may have, at this time, instilled the migratory instincts in each type of animal causing them to migrate to the ark.

How did Noah feed the animals and take care of their waste?

Even animals did not kill one another until the curse. The animals were originally vegetarians (Genesis 1:30) and probably remained that way until after the Flood. If "The wolf and the lamb shall feed together, the lion shall eat straw like the ox" (Isaiah 65:25) during the Millennium, then surely they could have lived peaceably together during the Flood. There were "rooms" in the ark (Genesis 6:14) which could have been nests or cages. God could also have caused the animals to hibernate, thus reducing the feeding and waste problems.[7] God had the ability to cause a deep sleep to fall upon Adam when he removed a rib to form Eve (Genesis 2:21). If medical science uses anesthesiology and induced comas, God surely knew about them before medical science discovered them.

---

[1] Stan Taylor, "The World That Perished," A video produced by Films for Christ, Mesa, AZ, 1977.

[2] Whitcomb and Morris, *The Genesis Flood*, 10.

[3] Taylor, "The World That Perished."

[4] Morris, *Biblical Basis*, 291-295.

[5] Arthur Hodges, III, "Dimensions of Noah's Ark," e-mail correspondence, November 7, 1997.

[6] Morris, *Biblical Basis*, 295-296.

[7] Whitcomb and Morris, *The Genesis Flood*, 70-75.

8 / Orogeny - The Formation Of Mountains

When were the mountains originally formed?

When God originally created the earth, it was covered with water. It was not until the third creative day that God said, "Let the waters under the heaven be gathered together unto one place, and let the dry land appear: and it was so" (Genesis 1:9). Since water seeks its own level, some form of elevated ground had to be created. Thus the first period of significant mountain formation probably occurred during the third day. The land was uplifted in order for the water to run off. From the above scripture, it appears that there was just one landmass at this time. This landmass, however, must have had considerable geographical features since the Bible says there were rivers (Genesis 2:10-14), high hills (Genesis 7:19) and mountains (Genesis 7:20). Since it did not rain until the Flood (Genesis 2:15), waters for the rivers were probably springs from the earth.

Where did all these flood waters go?

It is estimated that the water in the atmosphere today is only sufficient to cover the earth by a few inches. Thus only a small part of this water evaporated to form clouds and water vapor. The oceans, however, cover approximately 71% of the earth, and ocean basins are deeper than the continents are high. Some ocean basins are as much as 36,000 feet deep.[1] It has been estimated that if the earth were now smooth, water would cover the earth to a depth of 7500 feet, thus a flood is still here! All that was needed for

28

the floodwaters to decrease was for the mountains to rise and the ocean basins to lower. Psalm 104:6-9 probably refers to Noah's flood when it states, "...The waters stood above the mountains. At Your rebuke they fled; at the voice of Your thunder they hastened away. They went up over the mountains; they went down into the valleys, to the place which You founded for them. You have set a boundary that they may not pass over, that they may not return to cover the earth." Another way of saying this is that the mountains went up and the valleys went down.

Are the present mountains higher than the mountains before the Flood?

It would seem that the mountain ranges before the Flood were much lower and that the present mountain ranges of the world were formed during and following the Flood. God prepared a place for the receding waters by apparently lowering the ocean levels and raising the mountains, or making higher ones. As the mountains were raised and the ocean bottoms deepened, water naturally flowed from high ranges to the lower-lying ocean floor. Nothing ever seen by man in this present era, such as volcanoes, can account for the tremendous mountain structures.

Did the rotational speed of the earth change with the rising of the mountains?

Amazingly, with the formation of the mountains, the rotational speed of the earth did not change. If a man sits on a stool that can be rotated and someone else gives him a spin, he can slow down by stretching out his arms or speed up by tucking in his arms. Figure skaters and divers also use this principle of angular momentum. God had to carefully balance the mountain building with the valley formations both above and below water. We serve an awesome God "Who has measured the waters in the hollow of His hand, measured heaven with a span and calculated the dust of the earth in a measure? Weighed the mountains in scales and the hills in a balance?" (Isaiah 40:12).

---

[1] Morris, "Scripture and the Flood," 4.

9 / Geographical Distribution of the Animals

Why aren't there kangaroos near Mt. Ararat today?

Evolutionists try to claim that, since the continents were isolated, different species evolved suitable for the climates of the regions. Evolutionists sometimes use the mistaken idea that marsupials are only found in Australia to support the idea that they evolved there. Unfortunately for these evolutionists, live marsupials are found also in America and fossil marsupials are found even in Europe. Creationists believe that all the animals of the earth had their ancestors on the ark. That would mean that some sort of kangaroo was on the ark. How do we explain the fact that there are no kangaroos near Mount Ararat today? There were also lions in Palestine during Old Testament days, but none are there today. Both Samson (Judges 14:6) and David (I Samuel 17:36) killed lions.[1] No conservationists stopped them. These animals lived in a uniform climate before the Flood. After the Flood, they had to migrate to climates acceptable to them, just like birds migrate in the winter. Thus kangaroos probably became extinct in other parts of the world, either because of hunting or climate, leaving only Australia with naturally-occurring kangaroos.

How did kangaroos get to Australia?

There are several theories as to how the kangaroos got to Australia. One is the theory of continental drift that postulates that the continents broke off from one another, the portion breaking off to form Australia having the kangaroos. Even one of the best reconstructions of how Africa, South America, Europe and North America once fitted together has areas of overlap between these continents, and Central America is omitted altogether. Some reconstructions seem geometrically feasible, but they are impossible to explain by continental drift. Eastern Australia would have to be rotated to fit into eastern North America. Evolutionary geologists suppose that the plates move only two to eighteen centimeters per year. At this rate it would take 100 million years to form an ocean basin or a mountain range.[2] Some suggest that Genesis 10:25 (the earth was divided in the days of Peleg) gives justification to the continental drift idea. "Let the waters under the heaven be gathered together unto one place, and let the dry land appear: and it was so. And God called the dry land Earth..." (Genesis 1:9, 10). Since land is singular, it is possible that there was just one land mass when God caused dry land to appear on the third day. There are several problems however, with the continental drift theory. Why were kangaroos only living on the part of the earth that broke off? Also, by what mechanism did the continents drift away from each other? Energy considerations to move vast plates cause one to realize the whole idea of continental drift and plate tectonics is highly speculative and questionable. Claims of drift actually being measured today are also doubtful. The Bible does not directly speak for or against continental drift.

Another theory is that the kangaroos were transported to Australia on floating islands of entwined vegetation and trees torn loose from the banks of rivers and swept out to sea. These floating islands have been observed out of sight from land.[3] But why did the kangaroos float only to Australia?

A third theory is that humans could have carried the kangaroos in boats. It is uncertain exactly when men first began to burn and scrape out large logs to make dugout boats and canoes. Later craft include canoes of the North American Indians, the kayaks of the Eskimos, and the outrigger canoes of the Pacific Islanders. Jonah, who lived centuries before Christ, found a ship capable of sailing to Tarshish. This conjecture still does not answer questions such as where kangaroos originated and why aren't they there today?

The fourth theory is that there were intercontinental land bridges that aided the migration of animals. No one seems to have difficulty with the idea that men and animals once freely crossed the Bering Strait, which separates Asia and the Americas.[4] During the ice age, with more water held as ice at the poles, the sea level would have been lower, meaning that there were land bridges enabling dry-land passage from Europe most of the way to Australia. Also, there may have been major tectonic upheavals, accompanied by substantial rising and falling of sea floors. Ice melting from the ice age could have provided the water to cover the intercontinental land bridges that are now under water.[5] Did the kangaroos hop all the way to Australia? Probably so, but that does not mean one pair hopped all the way. The ancestors of present-day kangaroos may have established daughter populations in different parts of the world, which later became extinct, just like the lions in Palestine. How could creatures, that need a rain forest environment, trudge across millions of acres of parched desert on the way to where they now live? The answer is that the desert didn't exist then. It seems that the world is much drier now than it was in the early post-Flood centuries.[6]

When Krakatoa, a volcanic island that lies between Sumatra and Java, erupted in 1883, it caused one of the world's worst disasters. Much of the island was blown to bits and volcanic dust floated about this region for a year.[7] The island remnant remained lifeless for some years, but was eventually populated by insects, earth worms, birds, lizards, snakes and even a few mammals.[8] How these were able to cross the ocean we are not sure.

---

[1] Whitcomb and Morris, *The Genesis Flood*, 83.

[2] Ham, Snelling and Wieland, 43-45.

[3] Whitcomb and Morris, *The Genesis Flood*, 85.

[4] Ibid., 85-86.

[5] Ham, Snelling and Wieland, 198-200.

[6] Ibid., 201-203.

[7] The World Book Encyclopaedia, Vol. 11, 306.

[8] Ham, Snelling and Wieland, 198.

10 / Theories Of Evolution

What is the theory of evolution?

"The evolutionary system attempts to explain the origin, development, and meaning of all things in terms of natural laws and processes which operate today as they have in the past. No extraneous processes, requiring the special activity of an external agent, or creator, are permitted...Particles evolve into elements, elements into complex chemicals, complex chemicals into simple living systems, simple life forms into complex life, complex animal life into man" by means of innate properties.[1]

The theory of evolution makes the following assumptions which cannot be experimentally verified:

1. Non-living chemicals gave rise to living material.

2. Single-celled organisms gave rise to multi-celled organisms.

3. Invertebrates, which have no backbone or spinal column, gave rise to the vertebrates.

4. Fish gave rise to amphibians. For example, frogs begin life in water as tadpoles with gills and later develop lungs. They are cold-blooded and scaleless.

5. Amphibians evolved into reptiles. Reptiles, such as snakes, lizards, turtles, crocodiles and dinosaurs, are cold-blooded vertebrates, .

6. Reptiles formed birds and mammals.[2]

On what two foundations does the theory of evolution rest?

The twin pillars of evolutionary thought are chance and eons of time. Mathematical probabilities practically, if not totally, destroy pillar number one (the possibility of life originating by chance). (See chapter 19.) Therefore, time must be the hero of the plot. All evolutionists agree that unless there are immense periods of time, evolution hasn't got a ghost of a chance.[3] However, population growth statistics show that if man had been here millions of years, with even a conservative growth rate, we would all be crushed. Millions of people would be standing on top of each other. Thus pillar number two (eons of time) is also destroyed. (See chapter 20.) Yet evolutionists continue to make statements such as "However improbable we regard the origin of life, given enough time, it will almost certainly happen at least once...the impossible becomes possible, the possible becomes probable; and the probable becomes virtually certain. One only has to wait, time itself performs the miracle. Given enough time, things can evolve from the simple to the complex. Even though we cannot observe the process, even though we cannot

33

demonstrate it, even though we cannot prove or verify it, what we need is time."[4]

Why is it impossible to offer scientific proof of origins?

To prove something scientifically requires it to be observable and repeatable. A scientific investigator can neither observe nor repeat origins. A philosophy of origins can only be achieved by faith. Creation cannot be proved because creation is not taking place now. It is impossible to devise a scientific experiment to describe the creation process. Evolution cannot be proved because, if evolution is taking place today, it operates too slowly to be measurable. No one has ever observed that the changes in variations of organisms change the kinds into different higher kinds. Even if scientists could create life from non-life, or higher kinds from lower kinds, it would not prove that such changes took place in the past by random natural processes.[5]

What is the history of evolutionary theories?

1. Epicurius suggested that living things might have developed from simple forms. Paul encountered the Epicureans in Athens (Acts 17:18).

2. Aristotle (384-322 BC) believed in a gradual transition from the imperfect to the perfect and that man stood at the highest point of one long continuous ascent.

3. Lamarck (1744-1829) postulated that the formation of a new organ is the result of a new need which has arisen and continues to be felt by the organism. A worm wanting eyes could develop them or we could develop eyes in the back of our heads by merely wishing for them intensely enough. This postulate is unacceptable. Another postulate was that all changes occurring during a lifetime of an organism are transmitted to its offspring by the process of reproduction. This theory is known as the "inheritance of acquired characteristics." He would explain the long necks of giraffes as follows: Droughts on the plains of Africa required giraffes to stretch their necks to reach the few leaves and through hundreds of generations the giraffe acquired its long neck. This can be disproved in many ways. For example, a blacksmith's large muscular arm is not automatically passed on to his son. Also one scientist cut off the tails of twenty-two generations of mice and found that the tails of their descendants were no shorter than those of a similar group whose tails had not been cut off.

34

4. Darwin (1809-1882) popularized evolution with the publication of his book in 1859 *The Origin of the Species by Means of Natural Selection or the Preservation of Favored Races in the Struggle for Life.* His theory proposed that new species arise by the continued survival and reproduction of the individuals best fitted or adapted to the particular environment. This theory is known as the "Survival of the Fittest." Darwin would explain the long necks of giraffes by saying that only the longer-necked giraffes

survived and reproduced, thus having only long-necked giraffes. Shorter than average-necked giraffes died of starvation. A problem with this is how did short-necked animals in the same region survive? Also how did we get any long-necked giraffes in the first place?

5. DeVries in 1905 published his "Species and Varieties, Their Origin by Mutation." He made his observations in his garden with the evening primrose weed. Some of the plants were so different that he suggested that new species might arise by mutation.[6]

---

[1] Henry M. Morris, Scientific Creationism (Public School Edition), CLP Publishers, San Diego, CA, 1974, 10-11.

[2] Pavlu, 102.

[3] Wayne Jackson, *Creation*, 1.

[4] George Wald, "The Origin of Life," Scientific American, 45-53.

[5] Morris, *Scientific Creationism*, 4-6.

[6] Klotz, 38.

11 / Mutations

What are mutations?

Evolutionists propose that mutations are the mechanism whereby one species changes into another. "A mutation is assumed to be a real structural change in the hereditary material which makes the offspring different from its parents."[1] Mutations might be caused by errors in copying the DNA's genetic code.[2]

Why can't mutations produce evolutionary ascent?

1. Mutations are random and not directed.
2. Mutations are rare. Mutation rates in humans are approximately one per million gene duplications.[3]
3. Good mutations are extremely rare. 99.9% of all mutations are either harmful or lethal. For example the normal fly has an average life span of 40 days while those with a purple eye have a life span of 25 days.[4] The larger the mutation the more likely it will be harmful.
4. The net effect of all mutations is harmful.
5. Mutations affect and are affected by many genes. The probability of simultaneous good mutations in all the genes which control a given characteristic is practically zero.[5]

For mutations to produce evolutionary ascent would be like trying to go somewhere by taking 99.9 or more steps backward for every step forward.

What causes mutations?

Radiations, such as X-rays and ultra violet rays, temperature changes and mustard gases can all cause mutations. X-rays have produced the majority of mutations with the rate of mutations increasing with increasing dosage.[6]

The cause of mutations in nature is still not known.

---

[1] Pavlu, 153.

[2] Phillip E. Johnson, *Darwin on Trial*, InterVarsity Press, Downers Grove, IL, 1993, 37.

[3] Gary E. Parker, "Creation, Mutation and Variation," Impact Article No. 89, Institute for Creation Research, El Cajon, CA, November 1980.

[4] Klotz, 283.

[5] Morris, *Scientific Creationism*, 55-58.

[6] Klotz, 281.

12 / Vestigial Structures

What does the word "vestigial" mean?

The word vestigial means "remaining as a vestige (slight remnant or trace) of something that has disappeared or no longer is fully developed or useful."[1] In other words, vestigial means useless.

Vestigial organs are organs for which no function has been demonstrated. It is postulated that they once were functional but, in the course of evolutionary history, they lost their usefulness and gradually deteriorated.[2] At one time, evolutionists listed 180 structures and organs in man they thought were useless.[3]

What are some organs that were once considered vestigial?

1. *Pituitary gland.* The pituitary gland is now known to be so important for growth and other reasons that it is called the "master gland of the body."[4]

2. *Tonsils.* These are now found to help fight off infection.

3. *Coccyx.* Some pelvic muscles that are very essential to us are attached to it. A person can not sit comfortably without the coccyx and it protects the end of the spinal column. Some people incorrectly refer to this as a "tail." To be a tail a structure must be a caudal appendage (tail-like thing attached to something larger like an arm or leg or fin), with its own muscles, nerves and blood supply. The coccyx in the human is not a separate and distinct structure but is merely the bottom of the backbone.[5,6] After all, it does have to end someplace!

4. *Appendix.* The fact that it often causes trouble is not sufficient reason to say that it is useless. Other parts of the body, such as a throat, are also subject to infection but does that mean that a throat is vestigial? Some possible functions of the appendix are: 1) adding lubricating fluids to contents of intestines; 2) secreting small amounts of digestive juices; and 3) possibly manufacturing some white blood cells.[7] It is interesting to note that many monkeys have no appendix. If an appendix is vestigial, then evolutionists would have to say that a monkey has evolved further than man. Thus using a vestigial appendix as an argument for evolution is very superficial.[8]

What conclusions can be made regarding vestigial organs based on the evidence?

1. There is no evidence where useless organs deteriorate nor is there evidence that organs are in the process of evolving into something useful.[9]

2. The function of an organ may not yet have been discovered. Though the thyroid, pituitary gland and tonsils were all regarded as vestigial for many years, all have functions.

3. An organ's function may be taken over by another organ when it is removed. For example, the spleen produces red corpuscles before birth but in the adult it ordinarily does not. However, in cases of severe hemorrhage, the spleen may resume this function. Also, the human body often has two of the same organs as a margin of safety, such as the kidneys and lungs. One can live with only one kidney or one lung.[10]

4. The list of "useless" structures decreases as our store of knowledge increases. If any part functions during the developing person or animal, even though it may not in the adult, it cannot be called vestigial. The ignorance of scientists about the specific functions of structure does not prove they have no function.

What other creatures have structures that were once considered vestigial?

Whales have "hip bones" which support some of the internal organs and make places for the attachment of muscles. Though a whale is a mammal, it is strictly a figment of the imagination to claim that whales once walked. Evolutionists also thought the spur-like structures in snakes were once legs. It is now known that snakes use these structures for weapons and to secure traction during locomotion.[11]

---

[1] Thorndike-Barnhart, *Comprehensive Desk Dictionary*, Doubleday and Company, Inc., Garden City, NY, 1955.

[2] Klotz, 132.

[3] Morris, *Scientific Creationism*, 76.

[4] Cora Reno, *Evolution - Fact or Theory?* Moody Press, Chicago, IL, 1953, 46.

[5] Ibid., 68.

[6] Duane T. Gish, "Evolution and the Human Tail," Impact Article No. 117, Institute for Creation Research, El Cajon, CA, March 1983.

[7] Reno, 48.

[8] Klotz, 134.

[9] Pavlu, 20.

[10] Klotz., 132-133.

[11] Reno, 49.

13 / Recapitulation Theory Of Embryological Development

What does "ontogeny recapitulates phylogeny" mean?

This theory was popularized in 1866 by Ernst Haeckel, a German atheist. His idea spread widely with the pleasant sounding slogan: "Ontogeny Recapitulates Phylogeny" which means the biological development of the individual (ontogeny) repeats briefly (recapitulates) the evolutionary development of the individual (phylogeny). This theory speculates that in the development of the human embryo it passes through the evolutionary stages that were encountered in the evolution of man. It goes from protozoan to worm, to fish, to amphibian, to mammal and then to human. For over one hundred years evolutionists used this notion as one of the main "proofs" of evolution.[1] For example, evolutionists presented the so-called "gills" that human embryos supposedly possessed as evidence of man's evolution from amphibious ancestors. The human embryo does not have gills, but rather has folds of tissue that develop into parts of the tongue, lower jaw and neck.[2]

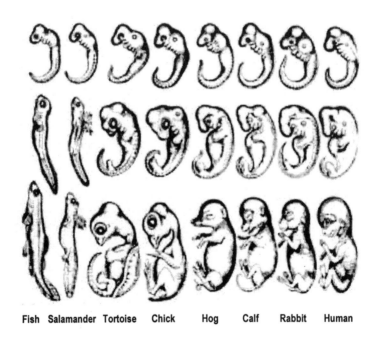

Fish Salamander Tortoise Chick Hog Calf Rabbit Human

What can be said about the drawings Haeckel made concerning embryos?

Haeckel made a series of diagrams of different embryos and demonstrated that they had certain characteristics in common. Haeckel later admitted that some of his drawings had been intentionally changed to make them fit the theory he was trying to prove.[3] Haeckel was a rogue.[4] If a student looked at the actual embryos instead of the diagrams, he would see more differences than the diagrams show.

What additional evidence falsifies the recapitulation theory?

Modern studies in molecular genetics have shown that the DNA for a man is not the DNA for a fish. The DNA for each kind is uniquely programmed to produce its own kind. This confirms I Corinthians 15:39, "All flesh is not the same flesh: but there is one kind of flesh of men, another flesh of beasts, another of fishes, and another of birds."

"Embryological studies have shown that there are many omissions, additions and inversions compared to the few parallels in the supposed evolutionary sequence."[5] Thus in no way could this represent an actual recapitulation.

How has the recapitulation theory been used as a pseudo-scientific justification for the plague of abortionism?

This concept of recapitulation is completely false and most competent evolutionists today know this. Unfortunately, this false notion is still propagated and is believed by millions of people today, leaving some bitter fruits.[6] The most recent application of the recapitulation theory has been as a pseudo-scientific justification for the plague of abortionism. If the fetus is not yet really a human, then what great harm is done if it is aborted? However, if the embryo is a human life with an eternal soul, then abortion is cruel, premeditated murder.[7] The abortionists say that the fetus is not "fully human," supposing that it is in the early stages of development from a "lower" form of life. What they fail to realize is that a human embryo develops into a human baby because that is what its genetic blueprint tells it to do.

What are some other devastating effects of evolution?

1. *Suicide.* The second leading cause of death among teenagers in the United States is suicide. One reason they commit suicide is because they have no purpose in life and have low self-esteem. They cannot answer the basic questions, such as "Where did I come from? Why am I here? or Where am I going?" Dr. Michael Girouard left a lucrative medical practice for two years in order to lecture on the subject of creation. He said that one young

man told him he was made in the image of an ape, who was made in the image of a bacteria, who was made in the image of a big bang, that was made in the image of nothing. He continued, "He came from nowhere, he was nothing and he was going no place, therefore he saw no reason to live." He asked the doctor, "Do you know what you do with an empty can of Vienna sausage? You throw it away." That is what the young man did. He committed suicide. But we know that you are not made in the image of an ape. You were made in the image of God. Your value is not based on what you do or your appearance or talents, but rather that you were created by the one who loves you enough that He died in your place. You are not junk![8]

2. *Communism.* Karl Marx wished to dedicate his book, *Das Kapital*, to Darwin. Many socialists and communists frequently refer to Darwinian ideas, such as class struggle.[9]

3. *Racism and Naziism.* See chapter 28.

4. *Destruction of faith.* See chapter 39.

5. *Gross immorality.* If God created man, He can set the rules. With evolution, every man can do what is right in his own eyes. If people are just products of time and chance, then there is no reason for treating men and women as objects of dignity and respect because they are no different from the animals from which they supposedly have evolved.[10] Evolutionists have no rules against fornication, adultery and homosexuality. No wonder sexually-transmitted diseases, such as AIDS, are rampant.

---

[1] Henry M. Morris, "The Heritage of the Recapitulation Theory," Impact Article No. 183, Institute for Creation Research, El Cajon, CA.

[2] Pavlu, 141-144.

[3] Reno, 65-66.

[4] Ken Ham, "The Smartest Man in America?" Back to Genesis Article No. 48, December 1992.

[5] Morris, *Scientific Creationism*, 1974, 77.

[6] Ham, Back to Genesis Article No. 48.

[7] Morris, Impact Article No. 183.

[8] Michael Girouard, Science, Scripture and Salvation Broadcast No. 214, Institute for Creation Research, El Cajon, CA, June 16, 1990.

[9] Morris, *Twilight of Evolution*, 18-19.

[10] Ken Ham, *The Lie - Evolution*, Master Books, P.O. Box 1606, El Cajon, CA, 1987, 88.

14 / Comparative Anatomy -The Study Of Similarities Between Different Kinds Of Animals

Why do evolutionists like to point out the similarities between man and animals?

"Comparative anatomy is the study of similarities between different kinds of animals. The term "homology" refers to the correspondence between the similar parts of different animals such as the arm of a man, the wing of a bird, the foreleg of a dog and the pectoral fin of a fish." Creationists view similarities as evidence of a common designer.[1] Evolutionists like to point out these similarities as their evidence for one species evolving into another. They especially like to point out the similarities between men and apes. It is true that there are many similarities, but to stress the similarities without even mentioning the differences is out of keeping with scientific objectivity.

How is man physically different from other mammals?

Some of the major differences are the following:

1. Permanent bipedal locomotion. Man is the only mammal that permanently walks on two feet. Whereas an ape can walk on two feet, they resort to all fours most of the time.

2. A nose with a prominent bridge and a well-developed, elongated tip. An ape more or less has a flat face with holes in it for nostrils.

3. A chin. The lower lip and jaw are the bottom of an ape's face with the absence of a chin.

4. Forward convexity of the spine, known as the lumbar curve. An ape's spine does not curve in at the back like a human's spine. If anything, it curves the other way.

5. Great toe on foot, not opposable to other toes like a thumb. All ten of our toes point in the same direction. The big toe of an ape is like the thumb of our hand.

6. A body which is relatively hairless. Hair on particular parts of the human body serves as a dry lubricant and odor absorber. Some people have more body hair than others do. Esau was hairy compared to his twin brother Jacob (Genesis 27:11). However, the hairiest human is relatively hairless compared to animals. Some have even called us naked or hairless apes.

7. A brain capable of learning, reasoning and talking. Whereas animals have learned some tricks and are able to imitate some sounds, the gap between the best-trained chimpanzees or birds and an average human being is immense. The English language has over one million words. No animal can come close to the reasoning power of a human.

8. A head that is balanced on top of the spinal column. Other animals have their heads at the end of the spinal column.

9. A very long period of postnatal growth. Some humans continue to grow into their twenties, whereas most animals reach maturity in a short time. Dogs are usually full-grown in a year and horses are winning the Kentucky Derby as three-year-olds. Perhaps God designed this long period of growth for humans because we have so much to learn before we become adults.

10. The highest total number of vertebrae.

11. The longest thumb in proportion to the length of the hand.

12. A different number of chromosomes.[2] "All flesh is not the same flesh, but there is one kind of flesh of men, another flesh of animals, another of fish, and another of birds" (I Corinthians 15:39).

One anatomy expert listed 312 characteristics that are found only in man.[3] These differences point out the need for transitional forms (missing links) which are completely missing. If man evolved from apes, where are the transitional forms, either living or in the fossil record, between man and apes? Most of what has been proposed as a transitional form has been proven to be a hoax or is so fragmentary that nothing can be definitely concluded.

43

How is man mentally and spiritually different from animals?

Man is suited for divine fellowship. Man's erect posture, upward gazing countenance, varied facial expressions, emotional feelings with brain and tongue for articulate speech are not shared by animals. When God made man, He made man in His own image and breathed on him and man became a living soul (Genesis 1:27, 2:7). Whereas animals have body (flesh) and spirit (life), they do not have a soul. Man has a portion that will live on forever. Man has a moral and intellectual nature that far exceeds any of the animals. Also man was created to have dominion over every living thing (Genesis 1:28). Jesus made it clear that man is better than the fowls of the air (Matthew 6:26) and sheep (Matthew 12:12). Man was created to rule, rather than evolve that way.

---

[1] Pavlu, 139-140.

[2] Klotz, 351.

[3] Taylor, *The Illustrated Origins Answer Book*, 234.

44

15 / Taxonomy - The Classification and Naming of Living Things

Who was the first taxonomist?

Adam was the first taxonomist in that he gave names to all cattle and to the fowl of the air and to every beast of the field (Genesis 2:20). What language he used and what he called them we do not know. Latin and Greek are used today in scientific classification because scholars in the past few centuries used these languages.

What are the seven chief groups in the scientific system of classification developed by Carolus Linnaeus?

Carolus Linnaeus (1707-1778) was a Swedish naturalist and botanist who established the modern scientific method of naming plants and animals. Linnaeus was born Karl von Linne in Sweden. He became famous as Carolus Linnaeus because he wrote his books in Latin.[1]

Every known animal and plant has a two-part name (binomial system of nomenclature). In this system, each living thing has a name with two parts. The first part is for the genus (group). The second part is for the species (kind).

Animals may have different common names in different parts of the world, but only one correct scientific name. For example, in different parts of the world the same animal may be called a puma, cougar, mountain lion, or panther, but its scientific name is Felis concolor. Felis is the genus and concolor the species.

Seven chief groups make up a system in scientific classification. The groups are 1) kingdom, 2) phylum, 3) class, 4) order, 5) family, 6) genus and 7) species. All animals belong to the animal kingdom, Animalia. All plants are members of the plant kingdom, Plantae. The animal kingdom may be divided into twenty or more phyla. For example, all animals with backbones belong to the phylum chordata.

Class members have more characteristics in common than do members of a phylum. For example, mammals, reptiles and birds all belong to the phylum chordata, but each belongs to a different class. Mammals have hair on their bodies and feed milk to their young. A platypus fits in this category, even though it swims and lays eggs. Reptiles have scales that cover their bodies and none feed milk to their young. Birds have feathers and also do not feed milk to their young.

Order consists of groups that are more alike than those in a class. For example, among the mammals Carnivora eat flesh, while Insectivora eat insects.

Family is made up of groups that are even more alike than those in the order are. For example, both wolves and cats eat flesh. Wolves, though, are in the family Canidae with long snouts and bush tails. Cats, however, belong in the family, Felidae, with short snouts and shorthaired tails.

Genus consists of very similar groups, but members of different groups usually cannot breed with one another. Both coyotes and timber wolves are in the genus Canis.

Species is the basic unit of scientific classification. Members of a species can breed with one another. The coyote is Canis latrans, and the timber wolf is Canis lupus.[2]

How does the ability to classify plants and animals confirm the Genesis account?

One of Linnaeus' main goals in systematizing the tremendous variety of living creatures was to equate his "species" category with the "kind" of Genesis. He believed in the "fixity of species," which means that variation could occur within the kind, but not from one kind to another kind.[3]

Evolutionists point to the similarities that allow living things to be classified as evidence for evolution, but actually this is one of the greatest weaknesses of the theory. The fact that there are numerous clear-cut gaps, which allow plants and animals to be classified, is evidence of the creation of separate "kinds." Creationists explain similarities as evidence of a common designer.[4]

---

[1] The World Book Encyclopedia, World Book-Childcraft International, Inc., Chicago, IL, 1979, Vol. 12, 295.

[2] The World Book Encyclopedia, 1979, Vol. 4, 500-502.

[3] Morris, *Men of Science*, 27-29.

[4] Pavlu, 144-145.

16 / Law of Biogenesis

How does the theory of evolution contradict the law of biogenesis?

The law of biogenesis is that life can only come from other life. It does not spring from non-living things.[1] This is what we observe and what the Bible teaches in Genesis 1, where various forms of life were created to "bring forth after their kind." Perhaps the most difficult problem that evolutionists face is the question of how self-replicating life systems could form from non-living, non-replicating systems.[2] Some evolutionists propose that in the beginning small inorganic molecules such as water, methane and ammonia, somehow by chance chemical reactions, formed amino acids. These amino acids combined to form proteins and eventually living cells.[3] The idea that living creatures can be produced naturally from non-living substances is called spontaneous generation. Never has this been observed, repeated or verified, and thus this idea is not scientific.

How close have scientists come to creating life?

Although scientists have synthesized amino acids, they have not even come close to creating life. This could be compared to a person who produced only three or four bricks and claimed to be near the completion of a fifty-story skyscraper.[4] Even if scientists could synthesize life, it would be the result of much planning, effort and time. This would not be a case of non-living matter spontaneously generating life.

What are some problems encountered in synthesizing life?

1. When chemists synthesize amino acids in a laboratory, they produce a 50:50 mixture of two different optical isomers. The two forms are known as L-amino acids and D-amino acids. The L and D indicate the direction in which polarized light is rotated when passed through the solutions. These isomers are mere images of one another and cannot be superimposed, just like a left hand cannot be superimposed on a right hand. Although chemically there is little difference, biologically the L-amino acids support life but the D-amino acids produce death. All living proteins are formed only from L-amino acids. Adding just one D-amino acid to a chain of L-amino acids can destroy the entire chain.[5, 6, 7]

2. The simplest living cells are far more complex than any amino acid or any machine ever invented by man. In chapter 19 I will show the mathematical improbability of life occurring by chance.

Is life just chemistry?

Life is made up of atoms, molecules, solutions and chemical reactions. However, when one dies, the atoms, molecules and solutions are still there. So whereas life involves chemistry, life is more than chemistry. The DNA molecule, which stores coded hereditary information,  is extremely complex. Many scientists are convinced that such cells could never have come into existence by pure, undirected chemistry. Mixing chemicals does not create DNA spirals or any intelligent code. Only DNA reproduces DNA.[8]

What two scientific experiments disproved the idea of the spontaneous generation of life?

1. Francisco Redi disproved the idea that maggots were spontaneously generated from decaying meat by placing samples of meat in three separate flasks. One flask was left open, another was covered with a porous cloth, and the third was covered with paper. All three meat samples decayed. The odor from the meat in the first two flasks attracted flies, which laid eggs on the meat in the open flask and on the porous cloth of the second flask, because they were not able to get to the meat. The paper covering provided an odor barrier, so the flies did not lay eggs on or near the third flask. The eggs hatched into maggots. Thus Redi showed that maggots came from eggs that the flies laid (life came from life) and not from spontaneous generation. This experiment destroyed the concept of spontaneous generation on a macroscopic level.

2. Even after Redi's experiments, some people thought that spontaneous generation could occur at the microscopic level. They observed that even a sterilized nutrient medium left exposed to the air soon was filled with microorganisms. Louis Pasteur demonstrated, however, that the organisms were introduced into the nutrient on air-borne dust particles. He established this by placing nutrient broth in a flask connected to the air only through a very narrow, sharply-curved tube. Airborne dust and bacteria were trapped on the interior surface of the curved tube and the broth remained sterile indefinitely, although the air had free access to it. Pasteur's work permanently settled the question of spontaneous generation. His experiment showed that sterilized broth would become cloudy with bacteria if dust could get to it, but would not be contaminated

if airborne bacteria could not get to it. Pasteur's experiment delivered a deathblow to the theory of spontaneous generation.[9]

What other ideas are proposed for the origin of life on earth?

Because of the problems encountered with life occurring by chance, some have suggested that life must have originated somewhere else in the universe and was then sent to the earth by some advanced extraterrestrial civilization.[10] These foolish speculations still do not explain the origin of life but merely transfer the origin of life to an unobservable part of the universe. How could life have survived the severe conditions encountered in space travel? It is so much more feasible to believe that God created life.

---

[1] Pavlu, 123.

[2] Morris, *Scientific Creationism,* 46.

[3] Gary E. Parker, "The Origin of Life on Earth," Creation Research Society Quarterly, September 1970, 97.

[4] Lawrence Richards, *It Couldn't Just Happen - Faith Building Evidences for Young People*, Word Publishing, Dallas, TX, 1989, 67.

[5] Duane T. Gish, "The Amino Acid Racemization Dating Method," Impact Article No. 23, Institute for Creation Research, El Cajon, CA, 1975.

[6] Taylor, *Illustrated Origins Answer Book*, 22-23.

[7] Brian Grantham, "My Favorite Evidence for Creation," Creation Ex Nihilo, Vol. 12, No. 1, December 1989-February 1990, 36.

[8] Taylor, *Illustrated Origins Answer Book*, 23.

[9] Pavlu, 122-126.

[10] Johnson, 110.

17 / Variation (Law of Kinds)

How does the theory of evolution contradict the "law of kinds"?

Genesis 1:24, 25 reads, "And God said, let the earth bring forth the living creature after his kind, cattle, and creeping thing, and beast of the earth after his kind: and it was so. And God made the beast of the earth after his kind, and cattle after their kind, and every thing that creepeth upon the earth after his kind: and God saw that it was good." The phrase *after his kind* is used at least ten times in the creation account (See Genesis 1:11, 12, 21, 24, 25). This applies to both the plant and the animal kingdoms. Specifically mentioned are grasses, herbs, trees, fishes, birds, beasts and creeping things. What this means is that pear trees produce pears and not bananas or monkeys. Cows have calves and horses beget colts. In other words, "like produces like." The theory of evolution contradicts the "law of kinds" by saying that one kind of creature evolved into another kind.

Why can't a dog breed with a cat?

Dogs produce puppies and cats have kittens. Dogs don't mate with cats and produce "dats." Although there is a tremendous variety of dogs and cats, they do not interbreed. Modern science has established that the code information in the DNA molecules does not allow reproduction to take place between a cat and a dog. The creator designed each kind to remain separate rather than blend.[1] Each kind of living creature has its own code (gene) and can reproduce only its kind. Variations are caused by variable forms of the same gene called allenes.[2] Abundant variety is possible within each kind but not between kinds.

What are some examples of variations in nature?

It is easy to observe tremendous variations in nature within the basic kinds of animals. For example, dogs range from tiny dogs, such as the Chihuahua and Pekinese, to huge dogs, such as the Saint Bernard or Great Dane. Selective breeding produces these different varieties. However if all the dogs were left together, after a few years the mongrels that would result would have less variety.

Small-scale biological changes or variations, such as color and size, are sometimes called micro-evolution. Micro-evolution does not produce new genetic information. God designed the genetic code with the ability to produce interesting variety within each kind. Creationists have no trouble with the concept of micro-evolution, except that the term is confused by some people with macro-evolution, which contradicts true science and the Bible. Macro-evolution would involve the production of new genetic information enabling large-scale biological changes (e.g. amphibian to reptile). Macro-evolution has never been observed.[3] Even within humans there is tremendous variation, but just one mankind. This will be discussed later in chapter 28 The Origin of Races.

Are the peppered moths an example of evolution?

There are black and white varieties of moths. Before the industrial revolution, tree trunks were relatively light colored and birds could easily spot the black moths, thus there were not many of them. The white moths blended in with the tree trunks and were not spotted as easily, thus there was a greater percentage of white moths. After the industrial revolution, soot from factories caused the tree trunks to blacken and the white moths were more easily spotted and eaten by birds. After this, the black moths began to be in greater percentage. No new kind of moths had evolved. The ratio of black moths to white moths was all that changed.[4]

What about the finches of the Galapagos Islands?

Darwin tried to use finches as evidence of evolution. While there is much variety in their bill size, coloration and other habits, they are all finches. This again illustrates the varieties possible within kinds.[5]

---

[1] Johnson, 110.

[2] Richards, 76-77.

[3] Taylor, *Illustrated Origins Answer Book*, 84.

[4] Pavlu, 154-155.

[5] Ibid., 151-152.

18 / Laws of Thermodynamics

How does the theory of evolution contradict the first law of thermodynamics?

True science has never discovered anything that contradicts the Word of God. Evolution, on the other hand, not only contradicts the Word of God, but it also contradicts known scientific laws, such as the laws of thermodynamics. The first law of thermodynamics deals with the conservation of matter and energy. This law tells us that nothing is now being created. Matter and energy can be converted from one form to another, but can neither be created nor destroyed.[1] This contradicts the idea that things are being created now (evolving). At the end of the sixth day, the Bible states that "...the heavens and the earth were finished and all the host of them. And on the seventh day God ended His work..." (Genesis 2:1, 2). Thus Creation is finished. God is preserving what He created in six days, but He no longer is creating anything except for occasional miracles.

How does the theory of evolution contradict the second law of thermodynamics?

The second law of thermodynamics shows that systems left to themselves go to a condition of greater disorder, probability and randomness.[2] Hurricanes do not build buildings. Explosions in junkyards do not build airplanes. Earthquakes do not create living systems. This law can be illustrated by dropping a handful of golf tees on an overhead projector. Even if the tees are in an ordered pattern when they are dropped, they will finish in a random pattern. Systems go from order to disorder. This contradicts evolutionary theory, which assumes that disordered particles eventually evolved to form ordered life. The second law of thermodynamics also contradicts the idea that a Big Bang Explosion could have produced an ordered universe. The golf tees can be arranged in a stick man pattern. Is the stick man a result of time and chance or the product of an intelligent designer? An audience may laugh at the word "intelligent," but they will clearly see that the stick man was purposely formed and not the result of an accident. For an ordered pattern to occur, there must be a designer and energy. Likewise, the orderliness of the universe and the complexity of living organisms confirm the work of a divine Creator. Our ordered universe could not have developed from chaos.

The second law of thermodynamics is also called a universal law of decay. Everything ultimately falls apart and disintegrates with time.[3] The sun is burning up, the rotation of the earth is slowing down and the magnetic field of the earth is decreasing.[4] Originally there was no disorder, decay, aging, suffering or death in the world that God created. Everything was very good until Adam sinned. It was then that the death originated. Genesis 3 describes some of the curses God placed on the earth when man sinned, such as difficulty bringing forth children, the growth of thorns and thistles, labor that produced sweat and, worst of all, death (unto dust shalt thou return). There are no exceptions to the second law of thermodynamics. Evolutionists try to point out that snowflakes forming, trees growing and embryos developing are exceptions to the second law. Snowflakes and other crystals form because of the sizes and shapes of atoms, ions and molecules that predetermine the shapes of the crystals. For example, if you dropped some marbles on a Chinese checkerboard, the marbles would take the pattern of the indentations on the board. This "disorder to order" is really not that at all, but is rather that the marbles are falling into a pre-designed order. Likewise, the order of a growing tree or a developing embryo has been pre-encoded into the cells of these systems.

---

[1] Morris, *Twilight of Evolution*, 32.

[2] Morris, *Scientific Creationism*, 25.

[3] Taylor, *Illustrated Origins Answer Book*, 7.

[4] Duane T. Gish, Impact Article No. 219.

19 / The Mathematical Improbability of Life Occurring by Chance

Why is the study of mathematical improbabilities important?

One of the strongest evidences for special creation is the mathematical improbability that the highly complex systems in the universe could have arisen by chance. Random processes generate disorder rather than order, and confusion instead of "information."

What are some easily explainable examples illustrating mathematical probabilities?

Mathematical probabilities of chance occurrences can be calculated precisely. For example, the chance of flipping a coin with heads up twice in a row is 1 in 4 (1 in 2 x 2). The coin could land head, tail or two tails or tail, head, or two heads. The probability of flipping ten heads in a row is 1 in 1028 (or 1 in 2 x 2 x 2 x 2 x 2 x 2 x 2 x 2 x 2 x 2 which can be expressed as 1 in 2^{10}). Thus, if you gambled and paid $5.00 for a chance to win $1000.00 if you could flip 10 heads in sequence, you would almost certainly lose. The probability of rolling five dice with each one showing a six is 1 in 7776 (1 in 6

x 6 x 6 x 6 x 6 which can be expressed as 1 in 6^{5}). Similar calculations can be made with cards or lottery numbers. Don't think that the casinos and state lotteries are not aware of the mathematics of probabilities. For example, the chance of winning the California Lottery is 1 in 18,009,640. Yet, in spite of the odds, people (often those who can't afford it) continue to waste their money.

Combination locks used on school lockers usually have 40 numbers. One usually turns the dial clockwise two times and stops at the first number. Then one turns it counter-clockwise one turn past the first number to the second number and then, lastly, clockwise to the third number. The chance of hitting the first number is 1 in 40. Likewise, the chance of hitting the second and third numbers also are each 1 in 40. The probability of hitting all three numbers by chance is 1 in 64000 (1 in 40 x 40 x 40). If it took a dishonest student one minute for each try, it might take him forty-four days to get the correct combination. By this time the security guards and the principal would have him arrested and expelled!

The chance of a young child, who doesn't know your telephone number, dialing your area code and telephone number correctly is 1 in 10,000,000,000 (1 in 10 billion or 1 in 10^{10}). Assuming the child tried ten hours per day and dialed once per minute, it would take approximately 50,000 years to reach you, which you would agree is improbable. From the above, you can see that the probability is calculated by the odds of an occurrence happening raised by the exponent which is the number of times it must happen in order. The chance of flipping a coin ten heads in a sequence was 1 in 2^{10}, because there were only two choices - heads or tails. But the chance of hitting ten telephone numbers correctly in order was 1 in 10^{10}, because there were ten choices each time (0, 1, 2, 3, 4, 5, 6, 7, 8 and 9).

The mathematics for calculating the probability of arranging a fixed number of items in order is somewhat different. Thus the possibility of arranging three flash cards in a predetermined sequence is 1 in 6 (3 x 2 x 1, or 3!) because the chance of getting the first card correct is 1 in 3, then the chance of getting the second card correct is 1 in 2. The last card is automatically correct if the first two are correct. The chance of arranging ten flash cards in a predetermined order is 1 in 3,628,800 (1 in 10!) while the chance of arranging 100 flash cards in order is 1 in 10^{158} (1 in 100!). Disorder is tremendously more probable than any kind of ordered system. The improbability of an ordered sequence increases as the number of components in the system increases.[1]

Why is the probability of life occurring by chance an impossibility as far as this universe is concerned?

A protein molecule is far more complex than arranging 100 flash cards in order. It has been estimated that the mathematical probability of the atoms of the simplest replicating protein molecule coming together in order by chance is 1 in 10^{450}. Astrophysicists estimate that there are no more than 10^{80} infinitesimal particles in the universe and that the age of the universe is no greater than 10^{18} seconds (30 billion years). If each particle can participate in a thousand billion (10^{12}) different events every second, then the greatest number of events that could happen in all the universe throughout its entire history is 10^{80} x 10^{18} x 10^{12} or 10^{110}. Any event with a probability of less than one chance in 10^{110} therefore cannot occur. Thus the probability of life occurring by chance is zero.[2]

A living cell is even far more complicated than the simplest replicating protein molecule. The chance of a single living cell spontaneously forming is 1 in $10^{40,000}$.[3] It is interesting to note the illustrations that various authors have used to illustrate the probability of even the simplest living cell forming by time and chance.

It is less than that of

1. Shaking for 1 billion years a gigantic box filled with wire, metal, plastic, etc., and forming a computer.[4]

2. A tornado passing through a junkyard and forming a Boeing 747.[5]

3. 10^{50} blind men simultaneously solving scrambled Rubic cubes.[6]

4. An explosion in a print shop producing an unabridged dictionary.[7]

What do the laws of probability and complexity prove beyond doubt?

Whenever one sees any real, ordered complexity in nature, particularly as found in living systems, he can be sure that this complexity was designed and did not occur by chance. Without a living God to create life, the laws of probability and complexity prove beyond doubt that life could never come into existence at all.[8]

---

[1] Henry M. Morris, "Probability and Order Versus Evolution," Impact Article No. 73, Institute for Creation Research, July 1979.

[2] Ibid.

[3] John W. Oiler, Jr., "A Theory in Crisis," Impact Article No. 180, Institute for Creation Research, El Cajon, CA, June 1988.

[4] Pavlu, 134.

[5] Johnson, 106.

[6] Jackson, *The Human Body Accident or Design?* Courier Publications, Post Office 55265, Stockton, CA, 1993, 3.

[7] Ibid.

[8] Morris, "Probability and Order Versus Evolution," Impact Article No. 73.

20 / Population Statistics Confirm the Bible

How many doublings of population were required to reach the present population on earth if one started with eight people?

A person really has only two choices regarding the origin and history of mankind. Either God was involved or God was not involved. Christians believe that God was involved and believe the biblical account of Creation and the history of man. Adam was the first man (I Corinthians 15:45) and his wife, who was made from his side (Genesis 2:22), was called Eve because she was the mother of all living (Genesis 3:20). Thus Adam and Eve were the original two people from which all others followed.

In the years that followed, the population apparently grew at a rapid rate because people lived to great ages and had many children. By the genealogies of Genesis 5:3, 6, 9, 12, 15, 18, 21, 25, 28, 32 and 7:6, one can calculate that the Flood occurred approximately 1656 years after the creation of Adam. Because of the wickedness of man, all of the people were killed during the Flood except for Noah, his wife, his three sons, Shem, Ham, Japheth and their three wives. The Scriptures plainly declare that only eight souls were saved (I Peter 3:20). Thus all people now on the earth sprang forth from these eight. It can be shown from the following table that it requires only thirty doublings of population to surpass the present population starting with eight persons.

Times Doubled Since Noah	Population	Times Doubled Since Noah	Population
0	8	16	524,288
1	16	17	1,048,576
2	32	18	2,097,152
3	64	19	4,194,304
4	128	20	8,388,608
5	256	21	16,777,216
6	512	22	33,554,432
7	1,024	23	67,108,864
8	2,048	24	134 million
9	4,096	25	268 million
10	8,192	26	536 million
11	16,384	27	1.2 billion
12	32,768	28	2.4 billion
13	65,536	29	4.8 billion
14	131,072	30	9.6 billion
15	262,144		

If Noah lived 4500 years ago, how often would the population have to double to reach the present population, and how does this rate of growth compare to the current population growth rate?

From genealogies (See Genesis 11:10, 12, 14, 16, 18, 20, 22, 24, 26; 21:5; 25:26; 29:35; 38:29; 49:10; Exodus 12:40; Ruth 4:12-22; I Kings 6:1; Matthew 1:1-16; and Luke 3:23-38) and history, one can show that Noah lived approximately 4500 years ago. By dividing 4500 years by 30 doublings one can show that to achieve our present population, starting with eight people 4500 years ago, would require the population to double approximately once every 150 years. This requires only an average growth rate of one-half percent per year. At present, the population growth rate is even greater than this. It is estimated that the population growth rate is two percent per year and the population is doubling in less than thirty years. I remember singing the missionary song, "A thousand million souls are dying, A thousand million souls for whom the Savior died." That song was written less than a hundred years ago when the population was one billion; now it is approaching six billion. This rapid population growth has occurred in spite of wars, diseases, birth control, abortions, etc. Thus, considering present growth rates, it is very reasonable to believe that the population of the earth sprang from eight people in 4500 years or less.

How many people would there be per square foot of land if the population had doubled 60 times since the time of Noah?

The alternative to believing the Bible account of Creation, the Flood, and the history of man is to believe that man evolved millions of years ago through an evolutionary process. But what would the population of the earth be if man had been on earth much longer? In looking at the above table, one can see that for each 10 doublings of population, the population increases about 1000 times. One could continue the above chart as follows:

Times Doubled Since Noah	Population
40	9,600 billion
50	9,600,000 billion
60	about ten billion billion

Thus, if we had just twice the number of doublings since the Flood (60 instead of 30) we would have about ten billion billion people on the earth. There would be even more than this, if there had not been a world-wide Flood, which reduced the population to eight. At an average doubling rate of 150 years per doubling, this would take only 9000 years or just twice as long as the biblical account.

What would have happened if man had been here millions of years as evolutionists propose?

Let us try to imagine what ten billion billion people would be like on the earth. The earth is basically spherical with a radius of 4000 miles. One can calculate the surface area to be equal to about 200,000,000 square miles. Since the earth's surface is about 70% covered by water, leaving about 30% land, there is approximately 60,000,000 square miles of land. With six billion people this is approximately 100 people per square mile. However, with ten billion billion people we would have over 100 billion people per square mile or over 4,000,000 people per square foot. In other words, in just 60 doublings of population (9000 years at an average rate of doubling each 150 years) we would have over one million people standing on top of the piece of paper you are reading. You would have been crushed a long time ago!

It is no wonder the population growth figures are causing such fear among people who are not knowledgeable about the Word of God. They foresee there will not be enough room or food for the people of the earth. That is why world leaders are attempting to limit population growth by so-called "family planning." We, who are Christians, need not fear because we have a hope that Jesus Christ is preparing a place for us and is coming to take us away to be with Him. (See I Thessalonians 4:13-18; I Corinthians 15:51-53.) After the

Tribulation and Millenium, there will be a great judgment where the earth will be burned up. In the days of Noah the earth perished by the Flood, but the future of the present earth is destruction by fire. We, however, look for new heavens and a new earth in which righteousness dwells (II Peter 3:9-13).

Even though the Lord does not want anyone to perish, because so many reject righteousness, there will be plenty of room for us in the place which God has prepared for us.

We can rely on the truths of the Word of God.

21 / Paleontology - The Study of Fossils

What are fossils?

Paleontology is the study of fossils. Fossils are the hardened remains or traces of plants and animals in the earth's crust. Literally billions of fossils have been discovered all over the world. They are usually found in sedimentary rocks that were deposited by water. Most fossils look like creatures that are living today, although some fossils show extinct forms of life.[1] Fossils of insects look like they died yesterday.

How are fossils formed?

Most often, the fossil-making process starts when a plant or animal gets buried rapidly.[2] Many fossils are formed by floods, which come without warning and carry enough sediment to bury living things. The weight of the sediment kills the creature, keeping its remains together. Creationists believe the Flood caused most fossils and that the animals and plants were all living at the same time, rather than in different geologic ages. Today when an animal dies, whether on land or sea, the bacteria in the body causes it to immediately begin to rot. Also scavengers, such as vultures, usually eat the carcass. These two agencies prevent the fossilization of most animals. For the carcass to be preserved, it must be buried deep enough so scavengers can't get to it and deep enough so oxygen, which bacteria need, is excluded. This implies that the animal must be buried quickly, or there will be nothing left to preserve. Ordinary deposition rates of sediment would not be adequate. The deposition of sediments had to have been thousands of times faster than the normal rates of deposition in order for a fossil to be preserved. It is reasonable to assume that the rapid deposition of sediments causing fossilization was caused by the worldwide Flood described by the Bible.

Why are fossils important?

Fossils furnish the only available records of the plants and animals of times past. Thus the fossil record is the only direct evidence for or against evolution.[3]

Why don't fossils reveal the shapes of soft parts of the body?

Usually the soft parts rot away and only the hard parts like teeth, bones and shells get preserved.[4] Since fossils usually are only the preservation of the hard parts, there is much we cannot tell from them, such as whether the animal was cold-blooded or warm-blooded.[5] Nor can we tell the shapes of soft parts, such as the nose, lips or hair. Most of what is seen regarding the shape or features of the reconstructions of men or animals is the figment of the imagination of the one doing the reconstruction. The person doing the reconstruction from fossils is often biased, due to preconceived ideas.

Why is the fossil record an embarrassment to evolutionists?

If evolution were true, we should find fossils of transitional forms. The fossil record though, gives little or no evidence of gradual change. The fossil record, rather than being a record of transformation, is a record of mass destruction, death and burial by water with its contained sediments. Fossils of trilobites, sponges, brachiopods, worms, jellyfish and other complex invertebrates are found in Cambrian rocks. The appearance of this great variety of complex creatures is so sudden that it is commonly referred to in geological literature as the "Cambrian explosion." If single-celled creatures gave rise to the vast array of complex invertebrates, the record of that evolution should be found somewhere in the Precambrian rocks. The intermediates between single-celled organisms and the complex invertebrates have not been found anywhere on this earth.[6]

The fossil record also fails to produce transitional forms between the major invertebrate types, between invertebrates and vertebrates and between the major fish classes.[7] For example, if fish evolved into amphibia, then transitional forms should show a slow, gradual change from fins into the feet and legs of the amphibians. Not a single transitional form has been found showing an intermediate stage between the fin and the foot. There is a basic difference in anatomy between fish and amphibians. In fish, the pelvic bones are small and loosely imbedded in muscle. There is no connection between the pelvic bones and the vertebral column. No connection is needed because the pelvic bones do not, and could not, support the weight of the body. In amphibians, however, the pelvic bones are very large and firmly attached to the vertebral column. An animal must have this type of anatomy to walk.[8] Also, if mammals (with a single lower jawbone and three bones in the ear) are supposed to have evolved from reptiles (with six bones in each half of the lower jaw and a single bone in the ear), where are the transitional forms, either living or in the fossil record? Not a single fossil creature has been found which represents an intermediate stage, such as one possessing three bones in the jaw and two bones in the ear.[9] Other differences between mammals and reptiles are the mode of reproduction, mammary glands, temperature regulation, hair and a different way of breathing. Mammals have a diaphragm. Reptiles have no diaphragm.[10]

The ability of insects, birds, mammals (bats) and reptiles (the pterosaurs now extinct) to fly was supposed to have evolved at four separate times. Not a single transitional form has been found in any series.[11] There are many differences between flying and non-flying creatures, such as flight muscles, hollow bones, etc. If evolution did occur, the developing wings would have been a distinct disadvantage to a reptile that had to compete with other animals designed for life on the ground. The reptilian jaws and teeth had to evolve into a toothless beak. What the animal was eating and how he chewed during this transition would be a mystery. This easily shows evolutionary scenario is an absurdity.

What the fossil record shows is not an unbroken sequence of gradual changes as proposed by evolution, but rather gaps. These gaps in the fossil record are a real embarrassment to anyone who believes in the theory of evolution. Different kinds of creatures simply appear in the rocks without any evidence of step-by-step changes between them. Charles Darwin, who authored the famous book on the Origin of Species in 1859, recognized the lack of fossil evidence for his theory but thought it would be found. Since Darwin wrote, millions of fossils have been uncovered, but there is still no evidence at all that a gradual change from simpler to more complex plants and animals actually took place. In actuality the fossil evidence is against Darwin's theory of evolution.[12] Most people are not aware that Darwin's strongest opponents were not clergymen, but fossil experts.[13]

What about archaeopteryx (ark-ee-'op-ter-iks)? ("old wing")[14]

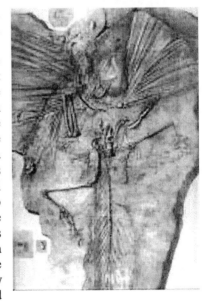

Evolutionists claim this fossil bird as an intermediary between reptiles and birds. It has many bird-like features, such as an avian wing, feathers like those of modern birds, perching feet and a wishbone. It also has some reptilian features, including claws on the wings, teeth and a long tail. Some birds today, however, such as the ostrich, the hoatzin of South America and the touraco of Africa, have claws on their wings. Since several fossil birds had teeth and many reptiles do not have teeth, teeth do not seem to be a good indicator as to whether a species is a bird or a reptile. Fossils of modern birds have also been found in the same rock layers as the archaeopteryx. How could it be the ancestor of modern bird types, if they existed side by side?[15] Bird fossils have also been found in some layers that evolutionists claim are about 75 million years earlier than where fossils of the archaeopteryx were found.[16] How could the archaeopteryx be the ancestor of creatures that supposedly lived 75 million years before? The teeth of the archaeopteryx were distinctly different from those of reptiles.[17] All the features of the archaeopteryx were fully formed and fully functional. There were not half scales and half feathers ("sceathers") or half legs and half wings ("lings"). One of the great arguments against evolution is that an animal could not have survived with only partially developed structures.[18] Thus the archaeopteryx was 100% bird and not a reptile-bird transition. It is simply another extinct bird that had teeth.[19]

---

[1] Richard B. Bliss, Gary E. Parker, Duane T. Gish, *Fossils: Key To the Present*, Creation Life Publishers, PO Box 15908, San Diego, CA, 1980, 4, 5.

[2] Taylor, *Illustrated Origins Answer Book*, 110.

[3] Pavlu, 63.

[4] Bliss, Parker, Gish, 6.

[5] Ibid., 16.

[6] Duane T. Gish, *Evolution, The Challenge of the Fossil Record*, Creation Life Publishers, El Cajon, CA, 1985, 54-56.

[7] Ibid., 69.

[8] Ibid., 72-73.

[9] Duane T. Gish, "The Mammal-like Reptiles," Impact Article No. 102, Institute for Creation Research, El Cajon, CA, December 1981.

[10] Gish, *Evolution*, 102.

[11] Ibid., 103.

[12] Richards, 84-85.

[13] Johnson, 45.

[14] Johnson, 80.

[15] Bliss, Parker, and Gish, 47-50.

[16] Taylor, *Illustrated Origins Answer Book*, 42, 103.

[17] Duane T. Gish, "Startling Discoveries Support Creation," Impact Article No. 171, Institute for Creation Research, El Cajon, CA, September 1987.

[18] Morris, *Biblical Basis*, 341-342.

[19] Morris, *Scientific Creationism*, 85.

22 / The Geologic Column

What is the geologic column?

Evolutionists believe there was once an age of fishes, later an age of reptiles and ultimately an age of mammals and man. Each age is supposed to have its own layer of rock, with the "older" layers supposedly at the bottom and the "younger" layers at the top.[1] The age of the layer is assigned using the index fossil method. The layer is dated by the fossils, and the fossils are dated by the layers. Thus one of the main evidences for evolution is the assumption of evolution.[2] This is circular reasoning. The geologic column is really just an idea rather than a column of rock.[3, 4] Creationists believe that the various ages and periods described in the geologic column never existed. The sedimentary layers were laid more or less continuously during truly massive flood conditions.

How dependable are index fossils in dating rocks?

The coelacanth (pronounced see-la-kanth)[5] was supposed to be a transitional form between fish and amphibians. It was thought to have limb-like characteristics on its fins[6] and to have lived during the Palezoic and Mesozoic eras. Supposedly, it became extinct after the Cretaceous period[7] some 70 million years ago.[8] Fossils of coelacanth were used as "index fossils" to date the strata in which they were found.[9] Evolutionists must have been embarrassed when fishermen started catching these fish in the 1930's.[10,11] The organs of the modern coelacanth show no signs of becoming adapted for land or of becoming an amphibian.[12] It is hard to believe that these fish became amphibians, if they are still the same as they supposedly were 70 million years ago. Creationists, of course, don't believe anything is 70 million years old. Needless to say, evolutionists no longer use the coelacanth as an index fossil.[13]

Where can the geologic column be found?

The complete "geologic column" does not exist anywhere in the world except in textbooks. Approximately 77% of the earth's surface on land and under the sea has seven or more of the ten strata systems missing.[14] Even the walls of the Grand Canyon include only about five of the ten strata systems. Also, in many areas, the strata are out of order because strata assigned to an "older" age in the column are found resting on top of strata assigned to a "younger" age.[15] This is not hard to explain with the cataclysmic effects of the Flood, involving giant landslides, tsunamis (huge sea waves caused by earthquakes or volcanic eruptions, also popularly, but inaccurately called tidal waves) and explosions, but it is difficult to explain with a uniformitarian concept. Likewise, the problem of fossils from different zones being mixed together is easily explained by a flood, but it is difficult to explain by evolution.

How do flood geologists explain the order of the fossils?

Dr. David Raup is Curator of Geology at Chicago's Field Museum of Natural History, which probably has the largest collection of fossils in the nation. Although he is an evolutionist, he has pointed out that Creationists have accepted the mistaken notion that the fossil record shows a detailed and orderly progression and that Creationists have gone to great lengths to accommodate this "fact" in flood geology. In other words, Raup is saying that flood geologists need not bother to work out a flood model for the order of the fossils, since there isn't any "order" to accommodate.[16] In some local situations, however, there appears to be an order of fossils with the marine invertebrate fossils in the lower strata and the mammal fossils in the higher strata.[17] Flood geologists have several reasons to explain this "usual" order of the deposition of fossils.

1. *Ecological Zonation.* Animals living at the lowest elevations would tend to be buried at the lowest elevations. It is natural to find the simplest marine invertebrates buried at the lowest level since they live in the deep ocean. Also animals would normally be buried with others living in the same region.

2. *Hydrodynamic Sorting.* Turbulent water is a highly effective "sorting" agent. Objects that are simpler in shape, such as trilobites which are extinct marine invertebrate animals with jointed legs and a segmented body divided by two furrows into three parts, and brachiopods which are marine animals with hinged upper and lower shells, tend to settle more rapidly than would objects of a complex geometry. Also these simpler organisms are of greater density, which would also increase their settling rate. The highly selective sorting action tends to deposit the simpler, more spherical organisms near the bottom of the sediments and to segregate

particles of similar sizes and shapes. This would give a superficial appearance of "evolution" of similar organisms in successively higher strata.

3. *Physical Mobility.* Mammals and bird fossils would be found, in general, at higher elevations than reptiles and amphibians, both because of their habitat and because of their greater mobility. Land vertebrates would be found at higher elevations because of their ability to run, thus escaping burial for a longer period of time. Few bird fossils are found, because only the exhausted were trapped and buried in sediment. Very few human fossils are found because men would have tried to escape by climbing to higher elevations, swimming and clinging to floating materials such as logs, thus almost entirely avoiding burial.[18, 19]

---

[1] David M. Raup, "Evolution and the Fossil Record," letter in Science 213 (July 17, 1981), p. 289, see also, by the same author "Geology and Creationism", Field Museum Bulletin 54 (March 1983), 16-25.

[2] Morris, *Scientific Creationism*, 1974, 136.

[3] Taylor, *Illustrated Origins Answer Book*, 40.

[4] Bliss, Parker, Gish, 14.

[5] Johnson, 71.

[6] Morris, *Scientific Creationism*, 82.

[7] Ibid., 89.

[8] Whitcomb and Morris, *The Genesis Flood*, 177.

[9] Morris, *Scientific Creationism*, 88.

[10] Ibid., 83.

[11] Klotz., 200-201.

[12] Johnson, 76-77.

[13] Morris, *Scientific Creationism*, 88.

[14] Steven A. Austin, "Ten Misconceptions About The Geologic Column," Impact Article No. 137, Institute for Creation Research, El Cajon, CA., November 1984.

[15] Taylor, 40.

[16] Ibid., 98.

[17] Morris, *Biblical Basis*, 361-362.

[18] Ibid., 329-330.

[19] Whitcomb and Morris, *The Genesis Flood*, 273-276.

23 / The Formation Of Coal and Oil

What is coal?

Coal is a sedimentary rock thought to be from accumulated, compacted and altered plants. Coal forms less than one percent of the sedimentary rocks. It is one of the strongest geological arguments for the reality of the great Flood of Noah's day.

What two theories are proposed for the formation of coal?

1. *The Autochthonous theory. (Swamp Theory)* Coal is supposed to have accumulated from trees and other vegetable matter which grew in swamps and later was buried. Compression under a vast weight of sediment is said to have resulted in the coal deposits. Because the accumulation of peat in swamps is a slow process, it is supposed that the coal beds required about 1000 years to form each inch of coal.[1]

2. *The Allochthonous theory. (Flood Theory).* Trees and other vegetable matter were rapidly transported by rivers into lakes and estuaries, where they were buried. Again, the pressure of the sediments would result in the coal deposits.

What evidence favors the flood theory for the formation of coal?

1. *Marine fossils in coal.* Often fossils such as fish, mollusks, and brachiopods are found in coal. The occurrence of marine animals with non-marine plants suggests mixing during transport, thus favoring the flood model.

2. *Polystrate fossils.* They have often been found in strata associated with coal. These upright tree trunks often penetrate tens of feet perpendicular to stratification. The sediments amassed in a short time to cover the tree before it could rot and fall down.[2]

3. *Cyclotherms.* Coal frequently occurs in a sequence of sedimentary strata called a *cyclotherm*. It is not unusual for *cyclotherms* to repeat many times with each cycle of deposition accumulated on a previous one. The Broken Arrow coal (Oklahoma), Croweburg coal (Missouri), Whitebrest coal (Iowa), Colchester No. 2 coal (Illinois), Coal Illa (Indiana), Schultztown coal (W. Kentucky), Princess No. 6 coal (E. Kentucky), and Lower Kittanning coal (Ohio and Pennsylvania) seem to form a single, vast seam of coal exceeding one hundred thousand square miles in area in the central and eastern United States. No modern swamp has an area anywhere near the size of this huge coal seam. *Cyclotherm* cycles can be explained better

by accumulation during successive advances and retreats of floodwaters rather than by the swamp theory.

4. *Boulders in coal.* Boulders have been found in coal beds all over the world. Even human skulls have been found. Boulders seem to have been entwined in the roots of trees and transported from distant areas.[3] The occurrence of boulders in coal also favors the flood model.

How does coal form?

Temperature, more than time, is the important factor in coal metamorphosis. Woody and other cellulosic materials have been converted to coal in a very short time in experiments performed by Dr. George R. Hill of the College of Mines and Mineral Industries of the University of Utah. In these experiments, the material was heated under high confining pressures and the properties of the products were similar to those found in anthracite and low volatile bituminous coals.[4]

The Noahic Flood could have transported and suddenly buried large amounts of vegetable material required for the formation of coal seams found in the earth today. The combination of burial and friction resulting from flooding and earth movements would have generated sufficiently high temperatures and pressures to convert this vegetable material into coal in a short time period.[5]

How does oil form?

Most geologists believe oil to be the converted remains of millions of trapped and buried marine animals. The Flood could have caused the trapping and burial and generated sufficient temperatures and pressures for rapid formation of oil.[6]

Experiments have been performed in which cow manure was converted to petroleum.[7] Several methods were used, each of which produced petroleum in a short time.[8] Garbage also has been converted to oil.[9]

---

[1] Steven A. Austin, "How Fast Can Coal Form?" Creation Ex Nihilo, Vol. 12, No. 1, December 1989- February 1990.

[2] John D. Morris, "What are Polystrate Fossils?" Back to Genesis Article No. 81, Institute for Creation Research, El Cajon, CA, September 1995.

[3] Stuart E. Nevins, "The Origin of Coal," Impact Article No. 41, Institute for Creation Research.

[4] Duane T. Gish, "Petroleum in Minutes Coal in Hours," Vol. 1, No. 4, 18.

[5] Ibid., 15, 16.

[6] Morris, Scientific Creationism, 109-110.

[7] Chemical and Engineering News, May 29, 1972, 14.

[8] Gish, Vol. 1, No. 4, 17.

[9] Larry Anderson, "Oil Made from Garbage," Science Digest, Vol. 74, July 1973, 77.

24 / Mount St. Helens

The top photo was taken on May 17, 1980 one day before the eruption. The bottom photo was taken on September 10, 1980 after the eruption. Both photos were taken from Johnston's Ridge six miles northwest of the volcano.

Why is the Mount St. Helens eruption so significant?

The eruption of Mount St. Helens in Washington State on May 18, 1980 could well be the most significant geologic event in the United States of the twentieth century. The total energy output of this eruption has been estimated to be the equivalent of 400 million tons of TNT or 20,000 Hiroshima-size atomic bombs.[1] This eruption, and the eruptions that followed, showed that great amounts of geologic work could be done rapidly by a catastrophe.[2] The Mount St. Helens eruptions produced changes, which geologists might otherwise assume required thousands of years. The Mount St. Helens eruptions were minor compared to the breaking up of the fountains of the deep during the Flood of Noah's day and the formation of the mountains which followed. Thus Mount St. Helens helps us on a small scale to imagine what the biblical Flood may have been like.[3]

What are some geologic effects that were brought about rapidly at Mount St. Helens?

1, *Stratification.* The eruption, and the flows which followed, resulted in layer formation that previously was thought to take thousands of years. Strata have formed up to 600 feet thick since 1980.

2. *Erosion.* A mudflow eroded a canyon system up to 140 feet deep in the Toutle River valley. This mechanism could have formed the Grand Canyon in a short period of time, rather than in the millions of years that evolutionists propose it took the Colorado River to form the canyon.

3. *Upright deposited logs.* The landslide from the eruption caused huge waves in nearby Spirit Lake. These waves stripped the adjacent forest and created huge log mats. Because of large root balls, some of the trees floated in an upright position, gradually sank to the bottom of the lake and were buried in sediment. This mechanism could have formed the petrified forests of Yellowstone National Park, rather than forests coming from many different eras.[4] The minerals in the water replace the organic material of the wood, forming a stone-like substance called petrified wood.[5]

4. *Peat layer.* The abrasive action of winds and waves caused much of the bark on the huge log mats in Spirit Lake to come off the trees and settle to the bottom of the lake. A layer of peat several inches thick has accumulated. All that is needed is burial and heating to transform this peat into coal. Geologists have previously thought that approximately 1000 years was required to form each inch of coal. The peat layer at Spirit Lake may be the first stage in the formation of coal.[6]

---

[1] Steven A. Austin, "Mount St. Helens and Catastrophism," Impact Article No. 157, Institute for Creation Research, El Cajon, CA.

[2] John D. Morris, "Mount St. Helens: Explosive Evidence for Creation," Notes presented at Institute of Creation Research Summer Institute, Northwestern College, St. Paul, MN, July 11, 1988.

[3] Austin, Impact Article No. 157.

[4] Ibid.

[5] Whitcomb and Morris, *Genesis Flood*, 166.

[6] Austin, Impact Article No. 157.

25 / Dinosaurs —"Terrible Lizards"

Why are dinosaurs so popular?

Children love to hear about dinosaurs, probably because of the "monster" appeal. They are bombarded daily through the newspapers, schools and other media with information regarding dinosaurs. Dinosaurs are probably used more than any other topic to brainwash children into accepting evolutionary ideas and rejecting the Bible as God's Word.[1] One of today's leading spokesman for evolution testified that he got his start by a childhood study of dinosaurs.[2] Dinosaurs have been used to sell everything from breakfast cereal to gasoline.[3] In 1989 the cover of Kellogg's Frosted Flakes advertised a free mystery dinosaur drawing disk inside the package. Pepsi Cola used drawings of dinosaurs on Diet Pepsi cartons and in 1992 included dinosaur trading cards that were similar to baseball trading cards. The movie *Jurassic Park* was produced at a cost of approximately sixty-five million dollars, which enabled the producers to develop the most sophisticated animated dinosaurs yet. This movie grossed nine hundred million dollars and the sales of Jurassic Park-related merchandise generated an additional one billion dollars.[4] In March 1993, a poster on dinosaurs was mailed free to seven million school children courtesy of McDonalds, which also handed out Jurassic Park mugs. In 1997, a sequel to the movie *Jurassic Park,* called *The Lost World,* was shown in the theaters. Despite not receiving rave reviews, it is still expected to be very successful. The most popular children's show on PBS television features "Barney," a big purple dinosaur. This show has 1.5 million viewers, which is more people than those who watch "Sesame Street."[5] In October 1997, the Chicago Field Museum of Natural History paid a record $8.4 million for a Tyrannosaurus Rex fossil.[6]

Did dinosaurs actually exist?

You may have grown up, as I did, in Missouri, the Show-Me-State, and perhaps wondered if dinosaurs actually existed or were just the product of the imaginations of evolutionists. All doubts were erased in my mind when I visited The Dinosaur National Monument Quarry near Jensen, Utah. It was there that paleontologist Earl Douglas discovered the world's greatest single deposit of fossil dinosaur bones in 1909. A quarry was built on the site, and it was designated a national monument in 1915. A year-round visitor center has been built over the quarry to protect the

73

fossilized dinosaur bones and skeletons. At the quarry, most of the bones are left in place. The fossilized remains of over 2,000 dinosaur bones are exposed as a permanent exhibit in the 200 foot-long sandstone wall. Visitors can watch paleontologists chip away the sandstone to expose the fossilized dinosaur bones. Visitors can also see the preparation laboratory where dinosaur fossils are cleaned and preserved. Nowhere else on earth can you see so many dinosaur bones still in their resting-place. The quarry has yielded a greater variety of species and a larger number of individual animals than any other single dinosaur site. Another unique feature of the quarry is the large number of juvenile dinosaurs found in it. You can look at dinosaur bones in many museums, but only at Dinosaur National Monument can you watch as experts actually unearth and study these fossils.

Great dinosaur graveyards have been found in various parts of the earth, giving evidence to the fact that dinosaurs really did exist.[7]

What does the term "dinosaur" mean, and why isn't the word "dinosaur" in the Bible?

Because the term dinosaur, which means "terrible lizard," was not coined until 1841 by Sir Richard Owen, the word is not in the King James version of the Bible which was translated several centuries before.[8]

What three words in the King James Version could refer to dinosaurs?

1. Dragons (Hebrew *tannim*) are mentioned at least twenty-five times in the Old Testament.[9] The first use is in Genesis 1:21. Some other references are Deuteronomy 32:33; Psalm 74:13, 14; 91:13; 148:7; Isaiah 27:1, 51:9; and Jeremiah 51:34. Also the word dragon appears a number of times in the book of Revelation.

2. Behemoth means a gigantic and powerful beast. Job 40:15-24 says "...he moves his tail like a cedar..." This description doesn't fit a hippopotamus or an elephant because neither of their tails is like a cedar. The behemoth was probably a land dinosaur such as a brontosaurus.

3. Leviathan was a type of dragon that was impractical to hook or harpoon (Job 41:2). It says of this creature, "He makes the deep to boil..." (Job 41:31). This leviathan was probably a marine dinosaur, such as plesiosaurus.[10]

What were dinosaurs like?

Besides the biblical descriptions, all we know about dinosaurs comes from fossilized bones and footprints. Whereas a biologist can learn about a modern animal by dissecting it, working down through skin, muscles, and organs to

74

the bones, a paleontologist must work in the opposite direction, starting with the bones. For example, the leg bones of a sauropod are stout, straight, and broad-ended. This enabled the legs to support a huge, heavy body, but didn't permit much flexibility in the joints. A modest amble of two or three miles an hour was probably the top speed of a full-grown sauropod.

Some things people can't determine from dinosaur bones and footprints are skin color and texture, flesh and other soft parts such as ears, eyes, and muscles that rot away. One also can't be sure whether the creature was warm-blooded or cold-blooded, nor can it be determined if the creature was vegetarian or meat eating. According to Genesis 1:29, 30, man and animals were originally vegetarian. It wasn't until after the Flood that man was given permission to eat meat. I don't think the teeth of men changed at that time, so one can't tell by the teeth if creatures were carnivorous.

When did dinosaurs live?

According to Genesis 1:20-25 dinosaurs must have been created on the fifth or sixth day. Evolutionists say that dinosaurs became extinct about 65 million years ago. Evolutionists place dinosaurs in the so-called Mesozoic era.[11] The Bible, however, says that "through one man sin entered the world, and death through sin, and thus death spread to all men..." (Romans 5:12), thus dinosaurs could not have died before man sinned. Creationists believe there was no death of any kind until Adam and Eve partook of the forbidden fruit. If this is the case, then men and dinosaurs coexisted. There are several lines of evidence for this that have been ignored or rejected by evolutionists. Two human skeletons were found a few miles away from the dinosaur fossils in the Dinosaur National Monument in the same Utah sandstone.[12]

Did Noah take dinosaurs on the ark?

If dinosaurs did not become extinct before the Flood, then Noah probably took representative ones on the ark. He could have taken young ones. The fact that Job describes behemoth and leviathan seems to indicate that dinosaurs lived for awhile after the Flood. Also dinosaur drawings, made by early tribal artists, have been found in Arizona, Siberia, Zimbabwe and elsewhere.[13] It is significant to know also, that in the ancient records and traditions of many nations, especially China, tales of dragons abound.

How did dinosaurs become extinct?

Creationists and evolutionists disagree as to when and how dinosaurs became extinct. An adult bull elephant can consume 300-600 pounds of fodder per day. Dinosaurs are estimated to have weighed over fifteen times as much as an elephant and probably consumed tons of food daily. Creationists believe that the lack of available food after the Flood was the cause for the extinction of dinosaurs.[14] It is likely that their final extinction came due to the changes in climate after the Flood. The leading evolutionary theory for the extinction of dinosaurs is that a large asteroid or comet struck the earth and the resulting dust killed the dinosaurs. A big question left unanswered then is "Why didn't this dust cloud kill the other animals also?"

Are any dinosaurs living today?

A picture was taken of a dinosaur-like creature dredged up near New Zealand in 1976.[15] In fact, some people think dinosaurs may still be alive today. It is also thought that many reptiles never stop growing until they die.[16] Could it be that some reptiles are miniature dinosaurs that never reach large sizes due to shortness of life? Some people feel that the Loch Ness monster in northern Scotland is a dinosaur. I have my doubts about this. Reports of a monster date back to the sixth century and have continued into the twentieth century. The monster has been described as a creature about thirty feet long.[17]

What conclusions can be made regarding the finding of dinosaur bones?

1. Fossil graveyards all over the world give evidence for a global flood, since fossils are produced by sudden burial.

2. Dinosaurs in Alaska give evidence for a warm uniform climate that existed before the Flood.

3. Change of climatic conditions after the Flood presents the best explanation for the "extinction" of dinosaurs.

4. Dinosaur fossils give credence and confirmation to the dragons, behemoths and leviathans of the Bible.

---

[1] Ken Ham, Dinosaurmania Strikes Again, Back to Genesis No. 55, July 1993.

[2] Henry M. Morris, "Dragons in Paradise," Impact Article No. 241, Institute for Creation Research, El Cajon, CA, July 1993.

[3] Paul S. Taylor, "Dinosaur Mania and Our Children," Impact Article No. 167, Institute for Creation Research, El Cajon, CA, May 1987.

[4] The Record, the newspaper serving San Joaquin County, Stockton, CA, September 12, 1994.

[5] Ham, Back to Genesis No. 55.

[6] The Associated Press, November 2, 1997.

[7] Morris, *Scientific Creationism*, 99.

[8] Morris, *Biblical Basis*, 351-352.

[9] Ibid., 351-352.

[10] Ibid., 357-358.

[11] Morris, Impact Article No. 241.

[12] Morris, *Biblical Basis*, 353.

[13] Ibid., 353.

[14] Pavlu, 36.

[15] Morris, *Biblical Basis*, 356.

[16] John D. Morris, "Dinosaurs," Science, Scripture and Salvation Broadcast No. 201, Institute for Creation Research, El Cajon, CA, March 17, 1990.

[17] The World Book Encylcopedia, 1979, Vol. 12, 366.

26 / Horse Evolution

Why is the horse evolution theory so important to evolutionists?

When asked to provide evidence of long-term evolution, most evolutionists turn to the fossil record. Fossil horses are frequently cited as evidence for practically every evolutionary principle that has ever been coined.

What is the historical background for horse evolution?

In 1841, the earliest so-called "horse" fossil was discovered near London. The scientist who unearthed it found a complete skull that looked like a fox's head with multiple back-teeth as in hoofed animals. He called it Hyracotherium because of the resemblance of this creature to the genus Hyrax (cony, daman). However, he saw no connection between it and the modern-day horse.

In 1874, another person, Kovalevsky, attempted to establish a link between this small fox-like creature, which he thought was 70 million years old, and the modern horse.[1]

In 1879, an American fossil expert, O. C. Marsh, and famous evolutionist Thomas Huxley collaborated for a public lecture that Huxley gave in New York. Marsh produced a schematic diagram, which attempted to show the transition of a small dog-like animal into a modern horse through a series of

intermediates. He published his evolutionary diagram in the American Journal of Science in 1879. Some years later, the American Museum of Natural History in the Smithsonian Institution in Washington, D.C. assembled a famous exhibit of these fossil horses, which were designed to show the horse family as evidence for evolution. This story of the horse family found its way into many other museums, publications and textbooks. The story and sketch typically illustrate the origin of a horse as follows. The earliest ancestor of the horse was called "Eohippus" (dawn horse) or "Hyacotherium." It was the size of a small dog and had four toes on each of the front feet and three on the rear, It had browsing teeth and lived about sixty million years ago. The next important ancestor "Mesohippus" (middle horse) was a slightly larger browser with three toes on the front feet and lived about thirty-five million years ago. After this came the larger "Merychippus" (cud-chewing horse), which had three toes, grazing teeth and lived about twenty million years ago. Next was the pony-sized "Pliohippus" which still had three toes, grazing teeth and lived about ten million years ago. These are supposed to have gradually lost their side toes. By about one million years ago, horses looked like modern horses under the name "Equus" with one toe per foot and grazing teeth. Tracing a line of descent from Hyracotherium to Equus would show apparent trends, such as the reduction of toe numbers, increase in the size of teeth, lengthening of the face and increase in body size.[2, 3, 4]

What are some serious problems with the theory of horse evolution?

1. The so-called "dawn horse" was not a horse at all. Instead, it was what scientists would call a hyrax, a rock badger or coney. In no way can Equus and Hyracotherium be considered the same "kind."[5] The "horse," on which the entire family tree of the horse rests, was not a horse at all.[6]

2. If the evolutionary scenario was true, you would expect to find the earliest horse fossils only in the lowest rock strata. Bones of the supposedly "earliest" horses have been found at or near the surface of the earth. Sometimes they have been found right next to modern horse fossils.

3. There are living horses with multiple toes.

4. Fossils of "horses" with three toes have been found at the same site as those with one toe,[7] showing they lived at the same time. In the January 1981 National Geographic on page 74, there is a picture of the foot of a so-called early horse, Pliohippus, and one of the modern Equus. Both were found at the same volcanic site in Nebraska.

5. There is no one site in the world where the evolutionary succession of the horse can be seen. Rather, the fossil fragments have been gathered from several continents on the assumption of evolutionary progress and then used to support the assumption. This is circular reasoning and does not qualify as objective science.

6. There are variations in the number of ribs within the imagined evolutionary progression. For example, the number of ribs in the supposedly "intermediate" stages of the horse varies from fifteen to nineteen, and then finally settles at eighteen.

7. Modern horses vary enormously in size. The largest horse today is the Clydesdale, while the smallest is the Fallabella, which stands at seventeen inches tall.[8] Both are members of the same species and neither has evolved from the other. The evolutionary assumption that the horse has grown progressively larger in size over millions of years has no value.

8. Some evolutionists admit the problem. Dr. Niles Eldredge, curator of the American Museum of Natural History, has said: "I admit that an awful lot of that (imaginary stories) has gotten into the textbooks as though it were true. For instance, the most famous example still on exhibit downstairs (in the American Museum) is the exhibit on horse evolution prepared perhaps fifty years ago. That has been presented as literal truth in textbook after textbook. Now I think that that is lamentable ..."[9]

How have evolutionists revised their theory regarding horse evolution?

As new fossils were discovered, such as mentioned above, evolutionists realized that the old model of horse evolution was a serious oversimplification. They still claim that the ancestors of the modern horse were roughly what that series showed and they were clear evidence that evolution had occurred. However, they now say that horse evolution didn't proceed in a straight line but was like a branching bush with no predetermined goal. Horse species were constantly branching off the "evolutionary tree" and evolving along various unrelated routes. We only have the impression of straight-line evolution because Equus is the only branch that survived.[10] Persons with an evolutionary bias go to great lengths to revise their theories rather than accept special Creation.

How can Creationists explain the sequence of horse fossils?

The horse series is a clever arrangement of the fossils on an evolutionary assumption. Contrary to what evolutionists say, these fossils have not been proven to be ancestors of the horse nor are they millions of years old. The dates evolutionists assign fossils are based on the assumed date of the geologic strata in which they are found.

Pavlu concludes, "the best evidence of horse evolution is the pictorial display in biology books, and even this is simply a projection of the evolutionists' imagination."[11]

---

[1] Peter Hastie, "What Happened to the Horse?," Creation Magazine, Sep.-Nov. 1995, Vol. 17, No. 4, p. 14-16, supplied by Answers in Genesis, http://www.christiananswers.net/q-aig/aig-c016.html

[2] Origins of the Horse, The World Book Encyclopedia, Vol. 9, 1979, 322.

[3] Kathleen Hunt, "Horse Evolution," The Talk, Origins Archive, Exploring the Creation/ Evolution Controversy, http://www.talkorigins.org/faqs/horses.html, January 4, 1995.

[4] John D. Morris, "What about the Horse Series?" Back to Genesis No. 63, Institute for Creation Research, El Cajon, CA, March, 1994.

[5] Hunt, Ibid.

[6] Duane T. Gish, "The Origin of Mammals," Impact Article No. 87, Institute for Creation Research, El Cajon, CA, September 1980.

[7] Gish, Ibid.

[8] Bliss, Parker, and Gish, 57.

[9] Niles Eldredge, as quoted in: Luther D. Sunderland, Darwin's Enigma: Fossils and Other Problems, fourth edition (revised and expanded), Master Book Publishers, Santee, CA, 1988, 78.

[10] Hunt, Ibid.

[11] Pavlu, 81.

27 / Human Origins

Why is physical anthropology so heavily influenced by subjective factors?

"Physical anthropology - the study of human origins - is a field that throughout history has been more heavily influenced by subjective factors than almost any other branch of respectable science."[1] The field is still begging for empirical confirmation. The pressure for confirmation is so strong that it has led to some spectacular frauds. The origin of mankind is often portrayed by evolutionists in chart form. They show a series of drawings of monkeys gradually becoming ape-men, and then man. They try to show from

the fossil record that prehistoric ancestors crawled upward from jungle darkness to become mankind. Dr. Derek Ager, former President of the British Geological Association, admitted "practically every evolutionary story he had learned as a student has now been debunked."[2] Such is the case with the stories of "prehistoric men."

What are some so-called "pre-historic men" that have been debunked?

1. *Ramapithecus*. The suffix "pithecus" means "ape." This fossil was found in India in 1932.[3] Time Life books claim that "Ramapithecus is now thought by some experts to be the oldest of man's ancestors in a direct line."[4] This hominid status is predicated upon a few teeth, some fragments of jaw and a palate unmistakably human in shape. This idea has been promoted through illustrations showing Ramapithecus routinely walking almost as upright as man, yet the fossil consisted of only a jaw and teeth. Most experts now concede that Ramapithecus was merely an extinct form of the orangutan. Orthodontists have shown that jaw angulation is an unreliable indicator of relative humanness and should not be the determining factor in deciding whether a jawbone is from a human or an ape. There is a large variation in the degree of angulation within both humans and apes.[5]

2. *Nebraska Man* (Hesperopithecus haroldcookii). Field geologist Harold Cooke sent a tooth to Henry Fairfield Osborn, director of the American Museum of Natural History, in order to determine the creature to which it had belonged. Osborn and some tooth specialists agreed that it was the first evidence of an anthropoid ape in the Western Hemisphere. They named him Hesperopithecus, ape of the west. It was known as the "Nebraska Man." After a period of 4½ years, the tooth was identified as the tooth of a wild pig. Prior to this exposé, it was felt that Nebraska man was halfway between Java Man and Neanderthal man.

3. *Piltdown Man*. In 1912, Charles Dawson and A.S. Woodward reported the discovery of an ape-like man in gravel near Piltdown, Sussex, England. What they found was a broken fossil skull and a jaw. In 1953, Kenneth Oakley proved that the skull and jaw did not belong to each other. The "Piltdown man" turned out to be a modern human skull and a jawbone of an ape. These bones had been chemically treated to make them look old and the teeth had been filed down to make them look worn. The forger had filed down the highly cusped molars of the orangutan jaw to give them the flat appearance of human cheek teeth. However, the filing had not been in the same plane and should have been obvious to experts who examined the specimen critically.[6, 7] Modern "science" was fooled for more than forty years. The whole thing was a hoax. British Museum officials protected these fossils from unfriendly inspection, allowing them to mold public opinion for over forty years.[8] This concealment of the evidence is a greater scandal than the original fraud.

4. *Peking Man (Sinanthropus)*. These fossils were found in the 1920's about twenty-five miles from Peking, China. The evidence consisted of skulls and teeth but almost no limb bones. After their discovery and description, the original bones were "lost" between 1941 and 1945. Some believe that the disappearance of the bones and the concealment of the human remains at the site may have been a cover-up by evolutionists trying to keep belief in Peking Man alive. Others believe it was just a twist of fate. One reconstruction of the skulls was named "Nellie." Most of the skulls had been bashed inward and fossil remains of humans and their tools were found at the site. Peking Man was probably an extinct variety of apes, once hunted and eaten in China by humans, who considered ape brains a delicacy.

5. *Australopithecus Afarensis*. Australopithecus literally means "southern ape." In Ethiopia in 1974, one Austalopithecus afarensis skeleton was discovered about 40% complete by Donald Johanson. He named it "Lucy" after a girl mentioned in a Beatles' song ("Lucy in the sky with diamonds..."). Everyone agrees that from the neck up "Lucy" was gorilla-like. Also, from the neck down nearly every feature was non-human.

"Lucy" was not a human ancestor. At best, she was a form of extinct ape; at worst, she was a combination of two or three species. She is still touted as one of the best "evidences" for human evolution.[9] Today these animals are no longer considered ancestors of man, but instead are a variety of extinct apes.[10]

6. *Java Man (pithecanthropus erectus)*. In the 1890's, Eugene Dubois went to the island of Java to discover the "missing link" and prove the theory of evolution. He found a skullcap, a thighbone and several teeth. This group of fossils was called "Java Man." DuBois only allowed a few privileged persons to see the fossils. Also, he did not reveal that he found the thighbone about a year later, about 45 feet away from the skullcap. Eugene Dubois was so obsessed with the findings of remains of ancient man that for thirty years he failed to make known the fact that close to the time he collected the famous Java Man fragments he also had found two human skulls (Wadjak skulls) in similar sediments. The "missing links" status of Java man would have suffered, if he had simultaneously revealed the Wadjak skulls.[11] Later research has shown the leg bone is almost unquestionably from a human and the skullcap is very ape-like. Dubois announced at the end of his life that the fossils were not the remains of an ape-man and that the skull belonged to a giant gibbon.[12]

7. *Neanderthal Man*. In 1856, workmen recovered a partial skeleton from a cave near Neander, Germany. Evolutionists seized upon Neanderthal as a missing link between apes and men. Neanderthal was reconstructed to show how he walked with a stooped stance and with a head set far forward. This appearance gave this man the characteristic ape-like look to lend support for Darwin's theory. It is now felt that Neanderthal Man was fully human, suffering from bone deformities that may have been caused by diseases.

Maurice Tillet was a professional wrestler from France, nicknamed "The Angel." Tillet's picture shows that his head and his face were very large.[13] His hands and feet were thickened and his torso broadened. One night Professor Carleton Coon and other members of the anthropological department of Harvard went to a wrestling match in Boston. After the match, they invited "The Angel" to Harvard to be measured scientifically. Tillet's measurements were identical to those of a Neanderthal Man![14] He was not a pre-historic man but he was a highly-educated and cultured individual. Apparently he suffered from an endocrine disorder which resulted from the overfunctioning of the pituitary gland. If one would have found his bones fossilized, he might have thought it was a skeleton of a Neanderthal man.

As academic dean for over twenty years in three different colleges, I have had the responsibility of ordering the caps and gowns for the graduates. In these years I have observed a tremendous variety of shapes and sizes of heads. One young man had a head so large that he cut his cap and sewed a piece of matching material in the gap. If years later the skulls of these graduates would be found and lined up, would someone would try to show an evolutionary pattern? In reality, all were living at the same time and not related. I also question the whole argument concerning brain size as an evidence for evolution. I would imagine there is great variation in the brain size of individuals living today.

Upon what evidence are the "origin of man" reconstructions based?

"Origin of man" reconstructions, which are commonly shown in books and museums, are based on very little evidence and a lot of imagination. Although millions of dollars have been spent, all the bones of the supposed ape links could be placed on a single table. On the basis of a few bone fragments or teeth, human-like faces have been added to ape bodies and vice versa. From bone fragments or teeth one cannot determine the shape of the nose, the eyes, the lips, or the ears, nor can one determine how much hair a person had or how much or where fat was deposited under the skin. By using the same skeleton and varying these features, an artist can produce all sorts of characters. Most of the time the artists have an evolutionary bias. Again, circular reasoning is employed. By assuming evolution is true, artists draw and sculpt primitive expressions and characteristics. They then display these reconstructions as evidence for evolution. These drawings and busts have had a powerful impact on the public imagination. With the "Nebraska Man," artists drew two people from one tooth. If they had found a skeleton, they might have made a yearbook!

Why don't we find more human fossils?

Prior to the Flood of Noah's day, people lived hundreds of years and had many children, thus there were probably millions of people living before the Flood. If most fossils were deposited during the Flood, one might ask "Why don't we find more human fossils?" First, consider that over 99% of the fossils are marine creatures or plants, and less than one percent are land animals. When land animals die in water, they usually bloat, float and come apart. It is difficult to trap and bury a bloated animal under water. Also, scavengers, seawater and bacterial action destroy everything.[15]

Why do people continue to claim to that they have found fossils of human ancestors?

One answer might be the love of money. There is much money available from the lecture circuit; from publications, such as National Geographic; and from organizations, such as the National Science Foundation. There is also the pride of life. Some fossil hunters have become famous and people seek their autographs. This is not the first time in history that people have falsely witnessed for money. The soldiers at the time of the resurrection of Christ received money to perpetuate the lie that the disciples stole away the body of Jesus while they slept. (See Matthew 28:12-14.)

In conclusion, there are no such things as ape-men. There are fossils of apes, and there are fossils of men, but nothing in between. Men have always been men and apes have always been apes. There are also some falsely-so-called scientists who have monkeyed around with the evidence.

---

[1] Johnson, 82.

[2] Duane T. Gish, *Evolution*, 81.

[3] Morris, *Scientific Creationism*, 172.

[4] Time Life Books, Early Man, Introductory Volume in the Life Nature Library.

[5] Taylor, *Illustrated Origins Answer Book*, 34.

[6] Roger Lewin, *Bones of Contention: Controversies in the Search for Human Origins*, Simon & Schuster, 1987, 188.

[7] Klotz, 365-369.

[8] Johnson, 82.

[9] John D. Morris, "Was 'Lucy' an Ape-man?" Back To Genesis, November 1989, Institute for Creation Research, P.O. Box 2667, El Cajon, CA.

[10] Malcom, *Ape-Men: Fact or Fallacy*, second edition, Bromley, Kent, England: Sovereign Publications, 1981.

[11] Harold Coffin, with Robert H. Brown, *Origin By Design*, Review and Herald Publishing Association, Washington, DC, 1983.

[12] Sylvia Baker, *Bone of Contention – Is Evolution True?* Creation Science Foundation Ltd., Queensland, Australia, 1986, 14.

[13] Klotz, 199.

[14] John Raymond Hand, "Why I Accept the Genesis Record," Back to the Bible Broadcast, Box 233, Lincoln, NE, 1959.

[15] John D. Morris, "Why Don't We Find More Human Fossils?" Back to Genesis, Institute for Creation Research, P.O. Box 2667, El Cajon, CA, January 1992.

28 / The Origin of Races

What terms does the Bible use instead of races?

Webster defines a race as "any of the different varieties of mankind, distinguished by form of hair, color of skin and eyes, stature, bodily proportions, etc. Many anthropologists now consider that there are only three primary major groups, the Caucasoid, Negroid, and Mongoloid."[1] The Bible does not refer to the word "race." Instead, it uses terms such as "tribe and tongue and people and nation" (Revelation 5:9).

We know from the Bible that Eve was the mother of all living (Genesis 3:20). Paul also declared that God "...has made from one blood every nation of men..." (Acts 17:26). The population grew rapidly until the time of Noah's Flood, when all people were destroyed except for eight persons (I Peter 3:20). Thus all human beings are descendants of Noah, his wife, his three sons, Shem, Ham and Japheth plus their three wives (Genesis 9:19).

How can prejudice influence ideas regarding the origin of races?

Prejudiced people, both religious and non-religious, have come up with a variety of theories for the origin of races. Some conclude that the Negroid race descended from Ham. Since Ham was cursed by Noah, they conjecture that this curse caused the skin to be dark and destined them to be servants (Genesis 9:25). This type of curse though, was usually for a maximum of ten generations, and more than ten generations have passed since this curse.

Early evolutionists were definitely racists. It is significant that Charles Darwin gave his book, *The Origin of Species by Natural Selection*, the provocative subtitle "The Preservation of Favored Races in the Struggle for Life." In one of his letters he was quoted as believing that the higher civilized races will eliminate many of the lower races. He believed that the gorilla and the Negro occupied evolutionary positions between the baboon and the Caucasians.[2] Thomas Huxley, the leading evolutionist of the last century said, "No rational man, cognizant of the facts, believes that the average negro is the equal, still less the superior, of the white man."[3] Henry Fairfield Osborn, professor of biology and zoology at Columbia University and president of the American Museum of Natural History's Board of Trustees, wrote, "The Negroid stock is even more ancient than the Caucasian and Mongolians, as may be proved by an examination not only of the brain, of the hair, of the bodily characteristics...but of the instincts, the intelligence. The standard of intelligence of the average adult Negro is similar to that of the eleven-year-old youth of the species Homo sapiens."[4] Thus they believed that the white man was a more advanced species, higher on the evolutionary ladder than the black man, with the black man being somewhere between an ape and the white man. This type of evolutionary thinking gave a pseudo-scientific justification for the selling into slavery of great numbers of blacks from Africa and, more recently, for the blatant racism of Nazi Germany. Hitler used the German word for evolution (Entwicklung) over and over again in his book *Mein Kampf* (My Struggle). Evolutionary ideas, quite undistinguished, lie at the basis of all that is worst in Hitler's book *Mein Kampf* and his public speeches.[5] He perhaps got the title for his book from Darwin's subtitle. Hitler criticized the Jews for bringing "Negroes into the Rhineland" with the aim of "ruining the white race."[6] Isn't it interesting how political pressure, rather than "scientific data," caused the "falsely so called scientists" to change their views?

How do evolutionists explain the origin of races?

Some have proposed that the fact that the skin will "suntan" or darken in the sun, could be the cause of the origin of dark skins in hot regions such as Africa. They postulate that each generation would get successively darker until eventually there would be a black race. This idea is based on the theory of the inheritance of acquired characteristics proposed by Lamarck. This theory proposes that a suntan acquired by an adult would be passed on to their babies. This is not so. The baby must also be in the sun to acquire the suntan. Lamarck's theory was proven to be false by an experiment in which the tails of mice were cut off, yet the baby mice continued to have tails. This experiment was conducted for twenty-two generations and the mice continued to have tails which were no shorter than a control group whose tails had not been cut off.[7] Thus, evolutionists do not have a satisfactory theory to explain the origin of the races.

How can Christians account for the origin of races of people with their different coloring and stature?

First of all, there was enough time before the Flood (over 1,600 years) to develop a large population with considerable variety. Noah's sons could have looked quite alike or quite different, as some brothers do. Fraternal twins, one black and one white, were born to Tom and Mandy Charnock in 1982 in Lancashire, England. The mother was part Nigerian and the father was English.[8] The wives of Noah's sons could have been totally different from each other and the resulting children could have been quite different. Some feel that the three major races of people are the direct lineage from the three sons of Noah. It seems that the sons of Noah did not form three races, but rather three streams of nations. Among each of these streams are both light-skinned and dark-skinned people.[9]

Thus, there needed to be a mechanism which caused the differences among people. The confusion of languages at Babel was probably the mechanism. Before this event "The whole earth was of one language and one speech" (Genesis 11:1). When the Lord confused their languages, the people could no longer understand each other and there was no alternative but to scatter. It is interesting to note that the Chinese language has characters which signify Noah's ark. In the genes of man, there is the possibility for a wide variation of physiological characteristics. Even the children of one set of parents can vary greatly in size, color and shape. It has been shown that in just one generation, one in sixteen children of parents with a particular gene combination could have the darkest skin color, while one in sixteen could have the lightest skin color.[10, 11] When people are free to marry whomever they wish, the dominant genes will result in people having a more or less uniform appearance. However, the confusion of languages at Babel effectively enforced segregation. Within these small inbreeding family groups, the recessive genes could quickly cause the distinct physical attributes which we call "racial characteristics."[12] These changes are termed "recessive Mendelian (genetic) characteristics." There is adequate genetic variation in the human genetic code to allow the development of a great number of distinctive characteristics in a few generations by the enforced segregation caused by the confusion of languages.

89

What chemical is responsible for the skin coloring of all humans?

There is only one kind of man—namely mankind. The cells that make up human bodies are similar for all humans, but are different from animals. The blood is also the same. The real divisions among men are not racial, physical or geographic, but linguistic. In actuality, there is only one race among men—the human race.[13] People of all nations are freely fertile and blood transfusions from one race to another are totally acceptable, confirming Acts 17:26. All humans have the same coloring that is caused by the chemical melanin, which is a brownish-black pigment found in skin and hair. This is the same chemical that is in bananas and causes them to turn brown when they ripen. People vary in the amount of melanin that results in the various shades of color of humans. An adult Negro has approximately 1/30 of an ounce of melanin. This is not much on which to base prejudice.[14]

---

[1] Webster's New World Dictionary.

[2] Paul G. Humber, "The Ascent of Racism," Impact Article No. 164, Institute for Creation Research, El Cajon, CA, February 1987.

[3] Morris, *Scientific Creationism*, 179-180.

[4] Humber.

[5] Morris, *Twilight of Evolution*, 18.

[6] Humber.

[7] Klotz, 25.

[8] Ham, Snelling and Wieland, 143-144.

[9] Morris, *Biblical Basis*, 432.

[10] Parker, Impact Article No. 89.

[11] Ham, Snelling and Wieland, 136-144.

[12] Morris, *Biblical Basis*, 432-434.

[13] Ibid., 429-430.

[14] Cosgrove, 49-50.

29 / Teleology - The Study of Design and Purpose in Nature

What simple examples can be used to illustrate the teleological principle?

William Palely (1743-1805), in his book *Natural Theology*, offered a strong argument for the truth of the Creator and His work. One example he gave was that of a man who crosses a field and finds two objects lying upon the ground, a watch and a stone. The finder immediately recognizes that nothing is unusual about the stone but that the watch is in a different category altogether. It displays planning and craftsmanship, beauty and usefulness. The watch clearly requires a watchmaker. It is not just the product of weathering, erosion, etc. Palely applied this common sense principle to earth's life forms. The complexity of living creatures far surpasses that of any watch ever constructed. Surely the design seen in living systems requires a designer. This recognition of intelligent planning in nature is known as the teleological argument. The universe appears to be carefully designed for the well-being of mankind from the atomic level up to the galaxies and beyond.[1, 2]

Gary Parker used a similar example of someone strolling down a creek bed, finding a pebble of some interesting shape and then spotting an Indian arrowhead. Whereas time, chance and erosion rounded the pebble, it is obvious that natural processes did not form the sharp angular chips of the arrowhead. Without seeing the Indian or his arrowhead-making process, he immediately recognizes evidence of design and purpose. Likewise, a haphazard mixture of aluminum, rubber, and gasoline does not make an object fly. There must be an airplane designer. An explosion in a brick factory does not produce a building. There must be architects and builders.[3]

B.J. Ranganathan uses the illustration of two people approaching a sandcastle on a beach. Both would agree that the castle exists and would probably agree on most of the things that could be observed, such as color, size and shape. They disagree, however, on how the castle originated. One believes that someone made it, while the other thinks it got there by the random forces of the wind and waves. Neither one can prove his position because neither one was there to see how the castle came into existence. Which explanation or faith is more reasonable?[4]

Ken Ham likes to use the example of Mount Rushmore when teaching in public schools. He asks the students how many of them believe the presidents' heads got there by millions of years of wind and water erosion. No one believes that, because they know the heads had to be carved according to a plan and design.[5] Designer Gutzon

Borglum worked fourteen years on the project, which was later finished by his son. The head of George Washington is sixty feet tall (five stories). The total monument is taller than the Great Pyramid of Egypt.[6]

What are some illustrations of the teleological principle in the plant and animal kingdom?

1. The walnut shell is designed to preserve the meat and to provide seed for future trees. The strength to weight ratio is phenomenal. The shell is a rigid dome shape with a heavily corrugated surface and is made of very lightweight material. Is the design of the walnut a product of evolution or is there a master designer?[7]

2. Even earthworms are especially designed creatures. They provide food for birds and they burrow through the ground, making an important contribution to the fertilization, aeration, and drainage of the soil. Could its valuable work come about through mutations or natural selection via its struggle for existence, as evolutionists would propose? Or is it better to recognize that the Creator designed and planned the earthworm in the beginning to be a servant of the plant world?[8]

3. The woodpecker is a bird that uses its strong chisel-like bill to bore holes in trees to search for insects. It must have some type of shock absorber system to prevent getting splitting headaches or beating its brains out. Most woodpeckers have their toes so arranged that they can cling to trees and climb up and down the trunks. Two toes are pointed forward and two are pointed backwards. They use their stiff tail as a support while they cling to a tree. Woodpeckers have long tongues coated with sticky saliva. They thrust out these tongues to spear insects and draw them out of hiding places. Actually, the woodpecker's tongue is not as long as it seems. It is attached to the hyoid, a structure of bone and elastic tissue. The hyoid, which loops around the skull, pushes the tongue out of the woodpecker's mouth when it is feeding. The woodpecker's bill, toes, tail, and tongue are

not products of time and chance, but rather, more evidence of the design of the great Creator.[9, 10]

4. The penguin is an unusual bird, in that its wings are short flippers designed for swimming, rather than for flying. Its feet, which are used for steering, are at the end of the body, rather than at the middle, as other birds' feet are. The long, thin feathers have fluffy tufts at their base, so that the wind and water can not penetrate. Under the feathers, there is a layer of blubber which serves as an insulator. The breeding pattern is extremely unusual. Penguins may walk as far as ninety miles to their breeding grounds. When the female lays its egg, she rolls it onto her feet and then covers it with a fold of skin to keep the egg from freezing. Then the male takes the egg onto his feet, covering it with his fold of skin. The female goes back to the water and feeds for nearly two months while the male stands with its back to blizzards and sleet waiting for the egg to hatch. Somehow the female returns just in time, picking out her own mate from thousands of penguins, to feed the chick half-digested fish. The male then goes to the sea to feed and get food. When he comes back he somehow recognizes his offspring and feeds the chick just in time. The parents take turns fishing and bringing food back. Could the amazing design and breeding patterns of the penguin be a product of evolution, or were they carefully planned and designed by the Creator?[11]

5. A wombat is a thickset, short-legged, tail-less, and somewhat badger-like marsupial that lives in Australia. The wombat's pouch opens backwards, not forwards, like the kangaroo. The wombat tunnels under the ground with burrows, sometimes one hundred feet long. If the wombat had a forward-facing pouch, the pouch would fill up with dirt, the young would die, and wombats would become extinct. An evolutionist guide in Australia stated evolution is really wonderful. Over millions of years it slowly turned the wombat's pouch around, so that it could burrow under the ground. You might ask what happened during the millions of years while it was turning around? No, the backward opening pouch is evidence of the Creator who designed things properly in the beginning.[12]

6. The giraffe is the tallest animal on earth, with heights reaching up to eighteen feet. The giraffe has several special design features. It has an extra large heart, in order to pump the blood all the way up to the head. The brain would be damaged if it did not receive enough blood. The brain would also be damaged if it received too much blood. This would be the case if the giraffe bent down to get a drink, if it were not for some other specially-designed features. The giraffe has some valves in the blood vessels of the neck, which close, preventing the blood from rushing to the brain. As added protection, there is also some spongy tissue below the brain. Do you think these special features, along with its long neck, all

evolved at the same time? It is much easier to believe that God created the entire giraffe.[13]

[1] Donald B. DeYoung, "Design in Nature, The Anthropic Principle," Impact Article No. 149, Institute for Creation Research, El Cajon, CA, November 1985.

[2] John D. Morris, "Did a Watchmaker Make the Watch?" Back to Genesis Article, Institute for Creation Research, El Cajon, CA, March 1990.

[3] Gary Parker, "Things That Are Made," Impact Article No. 62, Institute for Creation Research, El Cajon, CA, August 1978.

[4] B.G. Ranganathan, *Origins?* The Banner of Truth Trust, 3 Murrayfield Road, Edinburgh EH12 6EC, P.O. Box 621,Carlisle, PA, 1988.

[5] Ken Ham, "Watches and Wombats," Back to Genesis Article, Institute for Creation Research, El Cajon, CA, March 1990.

[6] The World Book Encyclopedia, 1979, Vol. 13, 734.

[7] Fred John Meldau, *Why We Believe in Creation Not Evolution*, Christian Victory Publishing Co., Denver, CO, 1959, 103.

[8] The World Book Encyclopedia, 1979, Vol. 6, 21.

[9] The World Book Encyclopedia, 1979, Vol. 21, 331.

[10] Richards, 103, 104, 123.

[11] Ibid., 104-105.

[12] Ham, "Watches and Wombats."

[13] Geoff Chapman, "The Giraffe," Creation Ex Nihilo, Volume 12 No. 1, December 1989-February 1990.

30 / Animal Migration

Why do animals migrate?

In a broad sense, migration means to go from one place to another. This chapter will not cover human migration but rather animal migration. By animal, we mean any living thing that is not a plant and, of course, not a human. Animal migration usually means a two-way journey that is a yearly round trip. The causes of animal migration have remained basically the same for thousands of years. Since wild animals can not control their surroundings as people can, they must move from area to area to find the physical conditions in which they can survive.[1] Migration has been observed among many types of animals, such as fish, reptiles, insects and birds.

The question regarding how an animal knows when and where to migrate is still not clearly known in many cases, even though there has been extensive research. Another question people ask is where did the animals get their migratory instincts?

What are some examples of animal migration?

1. *Pacific Salmon.* Pacific salmon are known as steelhead because of their gunmetal blue foreheads. They hatch from eggs in various streams and rivers of Washington, Oregon and Idaho. When they are about a year old, they drift down these rivers toward the Pacific Ocean where they feed and grow. One steelhead, which was tagged in Washington's Quinault River, was caught 3,210 miles from its home river, which is more than half of the distance to Japan.[2] When it is time to migrate back home to spawn, the steelhead swim back to the exact place where they were born. For example, those from the Salmon River in Idaho must go up the Columbia River, ignore the Deschutes River and many other tributaries, enter the Snake River and turn left up the Salmon river. The salmon must escape fishermen, traverse the ladders at eight huge hydroelectric dams, and fight their way up the raging torrents of the Salmon River until they reach the place of their birth.[3]

   How do the steelhead know how to find their way home? Many scientists believe they locate home streams by smell, since each river has a peculiar odor from its own soil and vegetation. They have performed experiments showing that many salmon with their noses plugged missed a crucial turn in a stream, while others did not.[4] How did the fish get this keen sense of smell and the determination to spawn at the place of their birth? Did this ability just evolve or did the Creator place it there?

2. *Monarch Butterflies.* This butterfly is known for its extraordinarily long migrations between central Mexico and the north-central United States and

Canada. The longest recorded flight for a tagged adult monarch was 1,800 miles. Many monarch butterflies spend their winters on fir trees high in central Mexico. In the spring they begin their journey northward. They lay their eggs on top of milkweeds to foster the next generation on the way. Several generations are born, breed and die before the final generation reaches as far as Canada. When autumn comes, after fueling up on flower nectar, that last generation flies all the way back to Mexico, though they have never been there before.[5] How do they know the route? How do they know the final destination, which is the same place Monarch butterflies spend the winter year after year? It is not a case of remembering, because this is their first trip.

3. *Birds.*Perhaps the most frequently observed and studied of all migration patterns is the migration of birds. Even Solomon mentioned "...the winter is past, the rain is over and gone. The flowers appear on the earth; the time of singing has come, and the voice of the turtledove is heard in our land." (Song of Solomon 2:11, 12). Jeremiah also wrote, "Even the stork in the heavens knows her appointed times; and the turtledove, the swift, and the swallow observe the time of their coming. But My people do not know the judgment of the LORD" (Jeremiah 8:7).

The arctic tern has perhaps the most remarkable known migration pattern of any animal. Each year it migrates from nesting grounds in the Arctic south to the Antarctic and back. The roundtrip journey is nearly 25,000 miles. Nesting occurs in the autumn on Arctic beaches or in areas of tundra.[6]

How is the migration of birds a miracle of energy?

The golden plover migrates from Alaska to Hawaii for the winter. Since there are no islands in route and the bird cannot swim, it cannot stop for rest or food. It is estimated that the bird makes 250,000 consecutive wing beats in the eighty-eight hours required to make this flight

of approximately 2,800 miles.[7] Birds have to have extraordinary stamina to travel the distances they do. Perhaps no creatures on earth are more athletic or as fit, not even Olympic athletes. The blackpoll warbler, which migrates from Nova Scotia to South America, loses half its weight in the four day 2,400 mile flight. Its fuel efficiency has been estimated as equal to 720,000 miles per gallon.[8]

How do the birds know how much energy is required to make their journey? Energy is stored in the form of fat but, to ensure the necessary flying capacity, the bird must be as light as possible and avoid excess weight. Also the use of fuel has to be as efficient as possible. If the bird flies too fast, it will waste too much energy overcoming air resistance. On the other hand, if it flies too slowly, it will consume too much energy just staying airborne. How does the bird know the optimum flying speed and also how to fly in V-formation to save energy? How does the bird know how much fat is necessary, the distance to fly and the economical rate of fuel consumption?[9] Since they don't go to aeronautical school, my answer is that God gave birds this knowledge.

How is the sense of direction which birds have a navigational miracle?

The bird's ability to find its way during migration is a great miracle, when one considers that the birds do not have a compass or map. Additionally, they have to navigate under constantly changing conditions, including sun position, wind direction and cloud cover. Without navigational methods, the vast majority of migrating birds would never reach their destination. If a golden plover were even slightly off course, it would miss Hawaii and end up dying in the open ocean. Since no species could survive such an overwhelming loss rate, any suggestion that evolution has played a part must be abandoned. Also, since many species fly solo, young birds can not learn the way by flying with their parents. Migratory birds have a God-given instinctive sense of direction that enables them to reach their destinations.[10]

What are some displacement experiments which have been carried out on migratory birds?

In one displacement experiment, two species of tern were shipped in different directions from their nesting places in the Tortugas Islands in the Gulf of Mexico and set free on the open sea. They were freed at distances ranging from 832 to 1,368 kilometers away from their nests over parts of the sea which were completely unfamiliar to them. Yet, within a few days, most of the terns returned almost directly to their eggs and young on the Tortugas Islands. In another displacement experiment, a Manx shearwater was taken from its nest on Skokholm Island in Wales and brought to Boston, Massachusetts. After a 5,000 kilometer nonstop transatlantic flight, which took twelve days, twelve hours and thirty-one minutes, it arrived back at its nest.[11] In another experiment, scientists took eighteen Laysan albatrosses from Midway atoll in the Pacific to Japan, the Philippines, the Mariana, the Marshall Islands, the Hawaiian Islands and Washington state. After releasing them, fourteen birds returned to Midway. The albatross from Washington traveled 3,200 miles, while the bird from the Philippines flew 4,120 miles.[12]

What are some disorientation experiments which have been carried out on migratory birds?

The navigational achievement of homing pigeons has been thoroughly researched and documented. On outward journeys, some birds were anesthetized and others were put in continuously-rotating cages. These birds were just as able to find their way home as the control birds. Homing pigeons have a special sense of geographical position and can determine their homeward course over long distances, even when all possible aids to orientation have been removed during the disorientation journey.[13]

How are migratory birds an evidence for Creation?

A bird's map sense is probably the most elusive and intriguing mystery in animal behavior.[14] There is no apparent way to avoid the tendency to drift off course over windswept stretches of ocean. This drift must be continually compensated for, as in a feedback system in control technology, otherwise the bird would lose energy by flying a longer route.[15] Scientists still don't know the nature of this instinct and which organ gives the bird this ability. Many ideas have been suggested to explain how birds navigate, such as the rising sun, star patterns, odors from pine trees, wind-generated low-frequency sound, and the magnetic field of the earth.[16] The best explanation though is that "the Creator equipped the birds with a precise 'autopilot,' which apparently is constantly measuring its geographical position and comparing the data with its individually 'programmed' destination. In this way, an economical, energy-saving, direct flight is guaranteed. How the navigational system works is known by no one today except the Creator, Who designed it."[17]

---

[1] The World Book Encyclopedia, 1979, Vol. 13, 449.

[2] Michael E. Long, "Secrets of Animal Navigation," National Geographic, June 1991, 70.

[3] Ibid., 76.

[4] Ibid., 89.

[5] Ibid., 72.

[6] "Arctic Tern," Microsoft® Encarta® 96 Encyclopedia, © 1993-1995 Microsoft Corporation.

[7] Werner Gitt, "The Flight of Migratory Birds," Impact Article No. 159, Institute for Creation Research, El Cajon, CA, September 1986.

[8] Long, 76.

[9] Gitt.

[10] Ibid.

[11] Ibid.

[12] Long, 76.

[13] Gitt.

[14] Long, 97.

[15] Gitt.

[16] Long, 78-79.

[17] Gitt.

31 / Geology - The Study Of The Earth

How does the earth illustrate purpose and design?

The earth, with all its unique features of rotation, tilt, revolution around the sun, air and water, is not a mere product of time and chance but is the work of the great Creator. The earth was formed as a place for people to live. "For thus says the LORD, Who created the heavens, Who is God, Who formed the earth and made it, Who has established it, Who did not create it in vain, Who formed it to be inhabited: "I am the LORD, and there is no other" (Isaiah 45:18). "The heaven, even the heavens, are the Lord's; but the earth He has given to the children of men" (Psalm 115:16). This makes me think of John 14:2, "I go to prepare a place for you." God is preparing another place for us to live eternally. The earth is the only planet on which life exists. Some type of life is found in every part of the earth. All other planets are covered with lifeless soil.[1] The earth also is the only planet covered with air, huge bodies of water and green vegetation for food, which were necessary if the earth was to be inhabited (Genesis 1:2-13).

Scientists and geologists can only study the present earth, which is composed of the three states: gas (mainly the atmosphere), liquid (mainly the hydrosphere), and solid (the lithosphere).

100

Why is the earth's atmosphere so important to us?

The atmosphere is the air that surrounds the earth. Without air, there could be no living plants or animals on the earth. People have stayed alive more than a month without food, and more than a week without water, but people can only stay alive a few minutes without air. How do the properties of the atmosphere benefit mankind?

It seems that God put the air in the firmament on the second day of Creation (Genesis 1:6-8). Air is about 1/5 oxygen and 4/5 nitrogen with traces of other gases. This mixture is perfect for life. If there were just a little more oxygen in the atmosphere, say 1/4 instead of 1/5, then the whole world would burst into flames.[2] If the atmospheric pressure were much lighter, or heavier, life would cease to exist on earth. If our atmosphere were much thinner, many of the millions of meteors, which now are burned up, would reach the surface of the earth causing death, destruction and fires.

Air rises when it is warmed. The air close to the surface of the earth is heated via light energy from the sun. The result is that the air near the surface of the earth is maintained at a temperature that is suitable for life. The movement of warm air from the surface of the earth creates wind, which carries pollutants away from our large population centers.[3]

Even the dust in the air is important. Without dust, neither drops of rain nor snowflakes would fall to the earth and clouds or fog would not form. Also, if it were not for dust in the atmosphere, the sky would be black. We owe the beautiful sunrises and sunsets to the presence of dust in the atmosphere.[4]

What is the function of each layer in the atmosphere?

The air surrounding the earth is classified into several layers.

1. The troposphere is the air layer that extends from the surface of the earth to where the temperature stops becoming lower. The troposphere is about six miles high over the North and South Poles and about ten miles over the equator. At the upper level of the troposphere, called the tropopause, the temperature is about −67 °F. Most of the air, moisture and dust of the atmosphere is in the troposphere. Clouds and weather are usually confined to this level.

2. The stratosphere extends from the troposphere to approximately thirty miles above the earth. The upper level of the stratosphere is called the stratopause and has a temperature of about 28 °F. Airline pilots prefer to fly in the stratosphere above the thunderstorms and snow found in the troposphere. The sun's rays warm the upper layer when they strike the ozone at this level. Many harmful ultraviolet rays are stopped by the ozone layer.

3. The mesosphere begins at about thirty miles above the earth and extends upward to about fifty miles. In this region the temperatures again decrease to about -135 °F at the top, which is called the mesopause.

4. The thermosphere begins about fifty miles above the earth and continues far into space. The air in the thermosphere is very thin and is fully exposed to the sun's radiation. It reaches over 2700 °F in the thermopause, which is approximately one hundred to three hundred miles high. The aurora borealis (northern lights) occurs in the thermosphere.

The ionosphere is a region of the atmosphere centered in the lower thermosphere. This region of electrically charged air particles makes it possible to have long-range radio communication without communication satellites. Radio waves of certain frequencies strike the ionosphere and are reflected back to the earth and can be received thousands of miles from their source.

The exosphere is the portion of the thermosphere above the ionosphere. Satellites or spacecraft encounter almost no resistance in this region because there is so little air. The exosphere is the outermost region of the earth's atmosphere.[5, 6]

What is known about the hydrosphere?

The hydrosphere covers about 70% of the earth's surface, with almost all of it in the ocean. Just off the oceans there is a continental shelf which ranges from less than a mile to two hundred miles out. It is rarely more than six hundred feet deep. Outside the continental shelf there is a continental slope which drops an average of 12,000 feet. Outside the continental slope are deep ocean basins, which can have deep underwater valleys and trenches. Some of these trenches are over 36,000 feet below sea level. In 1960, Jacques Picard and Don Walsh went to these depths in a specially-designed bathyscaph.[7] Also, in the deep ocean basins are sea mountains. The great fish must have taken Jonah down to some of these, for he said, "I went down to the bottoms of the mountains" (Jonah 2:6 KJV).

Why is water essential?

1. Man, animals and plants need it to live.
2. Plants use water to make food, which man eats.
3. Water makes soil by erosion and expansion.
4. Water provides temperature control.

How do the unusual properties of water benefit mankind?

Seventy percent of the earth is covered with water. This is a tremendous blessing to us because water has some very unusual and interesting properties. Some of these properties are as follows:

1, *High specific heat.* This is the amount of heat required to raise one gram of a substance one degree centigrade. The high specific heat of water prevents sudden increases or decreases in temperature between day and night. Water holds heat during the day and releases it during the night. This property reduces extreme temperature variations, especially in coastal regions and islands like England and Hawaii. Oceans store heat during the summer and release it during the winter. For example, the record low temperature in Juneau, Alaska, which is near the ocean, is warmer than the record low of St. Louis, Missouri, which is an inland city.

2. *High latent heat of vaporization.* This is the amount of heat required to evaporate one gram of a liquid. This property is important in cooling plants (transpiration) and animals (perspiration).

3. *High latent heat of fusion.* This is the amount of heat required to melt one gram of a solid. The high latent heat of fusion of ice prevents the snow on the mountains from melting too suddenly. Instead, we have water almost the whole year round. Farm wives use this property by putting a tub of water in the fruit cellar to prevent their jars of preserves from freezing, even on a bitterly cold night. The water must give up a lot of heat before it freezes. Oceans don't freeze in the winter because of this property.

4. *Expands when it freezes.* This property causes ice to float and lakes to freeze over at the top. The layer of ice then prevents the rest of the lake from freezing and enables the fish to survive. If water did not expand when freezing, the ice would sink to the bottom and the lakes would freeze solid and kill all the fish.

5. *Universal solvent.* More substances are soluble in water than in any other solvent. This helps in washing stains and dirt off our bodies and clothes. This property is also important because all of the necessary chemicals for life - proteins, starch, salts and sugars - can dissolve in water. Blood is mostly water and carries the chemicals of life throughout the body.

6. *Abundance and cheapness.* Seventy percent of the earth's surface area is covered with water, as oceans, lakes, seas and rivers. Also there are underground sources of water which are tapped by wells. Isn't it wonderful that to quench our thirst or to wash our hands, we don't have to buy something scarce costing $20.00 per gallon?

7. *Puts out fires.* Firemen use water to put out fires. Water, of course, doesn't burn and cools materials below the kindling temperature, partly because of its high specific heat and its high latent heat of vaporization.

8. *High boiling point for its molecular weight.* This prevents water from boiling away on hot days.[8, 9]

Why is gravity important to us?

God must have created the property of gravity on the second day to keep the water from falling off the earth. Also it would be needed later to keep the earth, the moon and other planets orbiting in the solar system. Gravity seems to be the only force that acts through space distances. God has given gravity the correct strength to enable people and animals to walk about freely on earth and not fly off into space. If gravity were stronger, birds and planes could not fly, clouds would lie on the ground, and standing up would be a difficult task. If gravity were less, the earth's atmosphere would escape into space, oceans would evaporate, and our orbiting distance from the sun would be greater, causing the earth to be too cold to be inhabited. In spite of the fact that Isaac Newton studied gravity three centuries ago, modern science still does not know what really causes the force of gravity.[10] The law of gravity and the second law of thermodynamics are two of the major laws that govern the universe.

What is known about the crust of the earth?

The lithosphere is the solid portion of the earth. The continents and the land beneath the oceans are a thin skin, called the earth's crust, that surrounds the main body of the earth. The crust, compared to the rest of the earth, has been likened to a very slightly-wrinkled apple skin over the apple. The thickness of the crust varies from about five miles under the oceans to about twenty miles under the continents. The highest land on earth is the top of Mount Everest in Asia, which is 29,028 feet above sea level. The lowest spot on earth is the Dead Sea in southwest Asia which is 1,299 feet below sea level. The deepest mineshaft is about two miles into the earth's crust, while the deepest oil and gas wells are about five miles deep. Thus, man has never yet penetrated the earth's crust. Part of the problem is that the temperature increases as one goes down. The temperature at the deepest part of the crust may reach 1600 °F, which is hot enough to melt rocks.

What are the three general categories of rocks found in the earth's surface?

1. Sedimentary rocks were formed at the surface of the earth, either by accumulation and cementation of mineral fragments or by precipitation of solid material from water. Sedimentary rocks sometimes contain fossil

remains of plants and animals. Limestone, sandstone and shale are examples of sedimentary rocks.

2. Igneous rocks form by the cooling and hardening of magma within the crust and of lava on the crust. Basalt, granite and pumice are examples of igneous rocks.

3. Metamorphic rocks are believed to be igneous and sedimentary rocks that have been changed by heat and pressure. For example, shale turns into slate, limestone into marble and bituminous coal into anthracite coal.

What is below the earth's crust?

The bottom of the crust is called the Mohoravicic discontinuity or Moho, which marks the boundary between the crusts and the inner parts of the earth.

Whereas scientists have not been able to go to the inside of the earth, they can study records of earthquakes. They believe that the inside of the earth is divided into three parts, the mantle, the outer core and the inner core. The mantle is believed to be a thick layer of solid rock below the crust going down about 1800 miles. The temperature of the mantle is believed to increase to about 4000 °F (2200 °C) where the mantle meets the next lower section, the outer core. Scientists believe the outer core is about 1400 miles thick and made of melted iron and nickel. The temperature of the outer core is believed to reach 9000 °F (5000 °C) in the deepest part. The inner core is ball-shaped and makes up the center of the earth which has a temperature of 9000 °F.[11]

---

[1] Jerry Bergman, "The Earth: Unique in the Universe," Impact Article No. 144, Institution for Creation Research, El Cajon, CA, June 1985.

[2] Richards, 32.

[3] Bergman, Impact Article No. 144.

[4] Meldau, 38-39.

[5] Bill W. Tillery, Physical Science Second Edition, William C. Brown Publishers, Dubuque, IA, 1993, 504-510.

[6] The World Book Encyclopedia, 1979, Vol. 1, 154-160.

[7] The World Book Encyclopedia, 1979, Vol. 2, 120.

[8] Klotz, 511-515.

[9] Bergman, Impact Article No. 144.

[10] Donald B. DeYoung, *Astronomy and the Bible*, 98.

[11] The World Book Encyclopedia, 1979, Vol. 6, 13-15.

32 / Astronomy and the Bible

What is the difference between astronomy and astrology?

The astronomical system was created for calendar keeping and for declaring God's glory (Psalm 19:1). To use it otherwise is idolatry. (See Deuteronomy 4:19, II Kings 23:5, Isaiah 47:13.)[1] God does not forbid astronomy, which is the science of the sun, moon, planets, stars and other heavenly bodies, their composition, motions, positions, distances and sizes. He does, however, forbid astrology, which is a false science that claims to interpret the influence of the stars and planets on persons and events. Astrology was developed by the Babylonians from their beliefs in the gods of nature.[2]

What was the origin of the universe?

There are only two choices for the origin of the universe. Either God was involved or He wasn't. "Through faith we understand that the worlds were framed by the word of God, so that things which are seen were not made of things which do appear" (Hebrews 11:3). "The 'Big Bang' is a popular, secular explanation for the origin of the universe. The process supposedly began with the explosion of a nugget, or 'kernel,' of mass energy, about fifteen billion years ago. As the energetic radiation spread outward, temperatures slowly cooled enough for hydrogen and helium atoms to form. About ten billion years ago, the first stars began to form from the cooling gas in the young universe. This star-forming process eventually gave rise to the Milky Way and other galaxies. When these initial stars had sufficiently aged, some of them became supernova explosions. The resulting star fragments later recombined into new stars to repeat the formation-disintegration process. Our sun is said to be a third-generation star, a relatively recent addition to the family of stars, and to have formed around five billion years ago. Other star fragments are thought to provide the material for planets and life forms, including people."[3]

What are some problems with the Big Bang Theory?

1. Where did the original concentration of mass energy come from?

2. What ignited the "Big Bang"?

3. How could an explosion produce order? An explosion should produce at best an outward spray of gas and radiation but certainly not planets, orbits or rotations.

4. Why is there not life in other portions of the universe? Life should have evolved everywhere in an evolving universe.

5. No natural way has been found to explain the formation of planets, stars and galaxies. The birth of a star has never been observed.

6. According to the Bible, the earth was formed before the sun, moon and stars.[4]

What was God's purpose for the heavenly bodies?

The sun, moon and stars were created on the fourth day (Genesis 1:14-19) for the following purposes:

1. *For rest.* God divided the day from the night so we would have a period of darkness to rest.

2. *For signs.* Solar eclipses are rare and very precisely specified in time. They have proven useful as "signs" in assigning dates of events in history. For example, Halley's comet can be predicted to orbit the sun at specific times.

3. *For seasons.* The earth is tilted 23.5° from a straight up position. This tilt and the earth's orbit around the sun cause the change in seasons. For example, the northern half of the earth tilts toward the sun in summer but away from the sun in the winter. If the earth were not tilted, we would not have seasons. The poles would lie in eternal twilight and huge continents of snow and ice would pile up in the polar regions, leaving most of the earth a dry desert. Life would soon be unable to exist on earth.

4. *For days.* All that was needed for day and night was for God to have a source of light and give the earth a spin. Each rotation of the earth causes a day and a night. If it spun slower, the days and nights would be longer. After the creation of the sun, this would mean much colder nights where life would freeze to death or burn up during the day from too much sun. God gave the earth the right spinning rate, and it has been spinning ever since, giving us our days and nights. The earth, with a circumference around the equator of approximately 24,000 miles, is rotating at the equator at a speed of approximately 1000 miles per hour. If it were not for gravity, we would be spun off from the earth like a piece of mud from off a bicycle tire.

5. *For years.* Each revolution of the earth around the sun causes a year. If the earth traveled faster than it does now, the centrifugal force would pull it further away from the sun and all life would cease to exist. If it traveled slower, the earth would move closer to the sun and all life would likewise perish. The earth's 365 day, 5-hour, 48-minute and 45.51-second round trip is accurate to a few thousandths of a second.[5]

What other purpose does the sun have?

The sun not only supplies daylight (See Genesis 1:16, Psalm 136:8), but it also provides energy for the continuance of life on earth. Only one billionth of the sun's energy output actually hits the earth.[6] In just one second the sun releases more energy than mankind has produced since Creation, including all the engines, power plants and bombs ever constructed. If the earth received just one percent less energy from the sun, it would soon be covered with ice. If the earth received just one percent more energy, the earth would soon be unbearably hot.

The amount of energy the earth receives from the sun depends on its distance from the sun, which in turn depends on the masses of the earth and sun and the speed of the earth in its orbit. The orbits can be determined by Kepler's laws of planetary motion. Ninety-three million miles of vacuum insulate us from the explosive sounds that would deafen us.[7] If the distance were less, we would be vaporized.

The sun is not just another star. Most stars put out a high proportion of lethal radiations like X-rays and gamma rays, but the sun provides us mainly with a life-supporting spectrum of energy. The sun is also unique in that its light and heat outputs are practically constant, while other stars fluctuate in output from 10% to 150,000%. Life on earth could not endure these extremes of radiation.[8]

Did these things just happen by accident? No. An Almighty and benevolent Creator specially designed the sun.

How does the earth's moon reveal purposeful design?

1. *Adequate night illumination.* Both Genesis 1:16 and Psalm 136:9 show that the moon was made to rule the night. The moon is almost 400 times smaller in diameter than the sun. However, the moon is almost 400 times closer, with the result that the sun and moon have the same apparent size in the sky. Among all the moons and planets in the solar system, this perfect match only occurs between the earth's sun and moon. The moon is just the right size to give a gentle, reflected light at night.

2. *An accurate time record.* The calendar month is based on the 29.5-day orbit of the moon about the earth. Eclipse data shows that the earth's

108

rotation of twenty-four hours has decreased by only 0.075 seconds in the past 3,000 years.

3. *Regular tides.* The gravitational force of the moon upon the earth causes tides on the earth, which cleanse the shorelines. Sometime in the future this may also provide an economical, non-polluting energy source. The tides are also an important component of the ocean currents. Without these currents, the oceans would stagnate along the seacoasts of the world, and the death of marine life, both animals and oxygen-producing plants, would soon follow. Our very existence depends upon the moon's tidal regulation of this intricate food supply. The moon is approximately one-fourth the size of the earth. If the moon was larger or if it was closer to the earth, huge tides would overflow the low lands and erode the mountains until the earth would be covered with water. If the moon was smaller or further away, marine life would be endangered in stagnant water.[9, 10]

One person said there is less chance of all these features happening by chance than for a blind person to drive through Los Angeles in rush-hour traffic and not have an accident.[11]

How was the moon formed?

Several natural origin theories that have been proposed are the following:

1. *Daughter or fission theory.* A moon-sized chunk of material broke loose from the earth.

2. *Wife or capture theory.* The moon had its own solar orbit but when it got near the earth it was captured by the earth's gravity and put into a permanent earth orbit.

3. *Sister or condensation theory.* The moon grew from dust and in the same region of space, as did the earth.

4. *Impact theory.* The Earth collided with a very large object, such as another planet, and the Moon formed from the ejected material.[12]

Angular momentum, force, orbital and frictional calculations have ruled out these theories.[13] The chemical makeup of the moon is also different from that of rocks on the earth indicating that the moon and the earth formed under different conditions.[14] Approximately twenty-five billion dollars was spent on the Apollo project. A total of twelve Apollo astronauts walked on the moon, taking thousands of photographs, conducting hundreds of experiments, and bringing back over 840 pounds of moon rocks.[15] In addition, they left instruments on the moon that continued to radio data back to the earth. According to them many questions were answered, but the big question that remains to be answered is how the moon formed in the first place. I could have saved them twenty-five billion dollars if they would have believed Genesis 1:14-19 which says God made the moon on the fourth day.

The moon remains in the sky as a faithful silent witness of God's creation (Psalm 89:37).

How do the stars give evidence of design and planning?

When David wrote, "The heavens declare the glory of God; and the firmament shows His handiwork" (Psalm 19:1), he probably was referring to the sun, moon, planets, and all the stars. I was never more impressed with the heavens than on one clear night in 1991, when I was in a rural area of southern Brazil where there were no lights. I'm not sure whether it was the Southern Hemisphere or the clarity of the night that caused the stars to shine so brightly. I have read that with an unaided eye a person could count 1,029 stars. With Galileo's telescope he could count 3,310 stars. Today, with the use of giant telescopes, it is estimated that there are 100 billion stars in our galaxy and that there may be 100 billion galaxies in the universe.[16] God knew before the invention of the telescope that there were too many to number.[17] God told Abraham that his seed would be like the stars of the sky (Genesis 15:5). Whereas man is not able to number the stars, God is so great that He not only knows the number but also calls them by their names! (Psalm 147:4-5). Every new space probe has reinforced our appreciation of God's creation. We know that the universe is not here by chance or accident. Instead, there is design and planning in every detail. Our universe is a cosmos (orderly universe), not a chaos (confused mass or mixture). The second law of thermodynamics tells us that our ordered universe could not have developed from chaos. The heavens were made for mankind to enjoy, but the earth was made to be inhabited. Space exploration continually shows how beautiful the earth is in comparison with other bodies in space.

Psalm 19:1-4 NIV

The heavens declare the glory of God; the skies proclaim the work of his hand. Day after day they pour forth speech, night after night they display knowledge There is no speech or language where their voice is not heard Their voice goes out unto all the earth, their words to the ends of the world.

So clearly does the Creation bear witness to the Creator that unbelievers are without excuse (Romans 1:20). A study of the galaxies shows us man's smallness and God's glory. No wonder David wrote, "When I consider Your heavens, the work of Your fingers, the moon and the stars, which You have ordained, What is man that You are mindful of him, and the son of man that You visit him? ..." (Psalm 8:3, 4a).

---

[1] DeYoung, *Astronomy and the Bible*, 102.

[2] Tillery, 343.

[3] DeYoung, *Astronomy and the Bible*, 86-87.

[4] Ibid., 89-90.

[5] Bergman, Impact Article No. 144.

[6] DeYoung, *Astronomy and the Bible*, 55.

[7] Ibid., 56.

[8] J. Timothy Unruh, "The Greater Light to Rule The Day," Impact Article No. 263, Institute for Creation Research, El Cajon, CA, May 1995.

[9] Donald, B. DeYoung, "The Moon: A Faithful Witness in the Sky," Impact Article No. 68, Institute for Creation Research, El Cajon, CA, February 1979.

[10] DeYoung, *Astronomy and the Bible*, 27-31.

[11] Richards, 32.

[12] DeYoung, *Astronomy and the Bible*, 27.

[13] DeYoung, Impact Article No. 68.

[14] Morris, *Scientific Creationism*, 31.

[15] http://www.seds.org/nineplanets/nineplanets/luna.html

[16] DeYoung, *Astronomy and the Bible*, 57.

[17] Richards, 8-9.

33 / The Search for Life on Other Planets

Has life been found other than on the earth?

Billions of dollars have been spent and are being spent on a search for life in space. The first place outside of earth to look for life would be the solar system. The solar system consists of the sun, nine planets and their moons, plus asteroids and comets. The inner planets, which are those closest to the sun, are Mercury, Venus, Earth and Mars. These planets are solid rock. During the 1960's, United States and Russian space missions were sent to Venus. They revealed that environmental conditions on the surface of Venus could not support life as we know it. In 1976, two United States space probes, Viking I and Viking II, landed on Mars. Experiments indicated chemical activity in Martian soil, but a highly sensitive chemical detector failed to detect any organic compounds. The outer planets are those farthest from the sun. These planets are Jupiter, Saturn, Uranus, Neptune and Pluto.

The following information shows why life does not, and could not, exist on any planet besides earth.

1. The estimated temperature on Mercury ranges from a high of 930 °F on the light side to a low of -350 °F on the dark side.

2. Venus rotates only once every 243 days. The surface temperature is about 870 °F. The atmosphere consists mainly of carbon dioxide, and the atmospheric pressure is estimated to be one hundred times the atmospheric pressure on earth. There are clouds of sulfuric acid. No living thing that we know could survive on Venus. None of the space probes sent to Venus have operated for longer than an hour before ceasing to function.

3. Mars has no water vapor. The average temperature is 63 °F below zero. Water in liquid form couldn't exist because the atmospheric pressure is so low. Mars is dry, cold, and desert-like, with no possibility for life to exist. The Viking mission confirmed no life on Mars. The idea that men from Mars exist is merely science fiction. The surface of Mars was photographed on the Mars Pathfinder Mission in 1997.

4. Jupiter is the largest of the planets, but it is not a solid planet. It is a mixture of ices and vapors.

5. Saturn is readily identified by its beautiful system of rings. It is less dense than water. Winds rage at 1000 miles per hour.

6. The temperature of Uranus is -350 °F and Neptune is even colder at -360 °F.

7. Pluto is the farthest planet from the sun. Probably it is smaller than our moon and is composed of frozen methane gas.[1,2]

In summary, there is no evidence of life on any planet other than earth. "The heaven, even the heavens, are the Lord's; But the earth He has given to the children of men" (Psalm 115:16).

The harsh environments on the other planets of our solar system make it highly unlikely that life could exist on any of them. Even if life existed on other planets, that would still not answer the question of how life originated. It is also highly improbable that life could survive the intense cold, extreme dryness, lack of oxygen and intense radiation in its journey through space.[3]

Thus exobiologists have to look outside of our solar system for life. But, with our present technology, it is impossible to even detect the presence of other planets, much less to search for life on them. Therefore, the only way they could learn of life on other planets would be to receive communication signals from other intelligent beings capable of communicating. Some have begun to listen.[4] Good luck!

---

[1] DeYoung, *Astronomy and the Bible*, 38-41.

[2] Tillery, 371-385.

[3] Pavlu, 126.

[4] The World Book Encyclopedia, 1979, Vol. 12, 245-246.

34 / The Human Body

How does the human body give evidence for Creation?

Evolutionists have no mechanism for the evolution of complex organs, such as the eyes, ears and lungs. God designed eyes to see, ears to hear and lungs to breathe.[1] The more one studies the human body the more one appreciates the statement of David "...I am fearfully and wonderfully made: marvelous are thy works; and that my soul knoweth right well" (Psalm 139:14). This is true whether one studies the human cell on a microscopic level, the individual organs, the various systems of the body or the body as a whole.

In no way could all the details of the human body be covered in one book, nor could you learn all about the human body in a lifetime. The medical field has specialists for various areas of the body, such as ophthalmologists for the eye, dermatologists for the skin and cardiologists for the heart. They also have specialists for different stages of life, such as pediatricians for children and geriatricians for old age. Even those in these specialized fields would admit that they don't know everything regarding their specialties.

How does the human body illustrate the teleological principle?

1. *The Eye.* Our most important sense organ is the eye. With our eyes we can see very bright items, such as the sun, and very dim items with amazingly low levels of light energy. We can also discern shapes and colors. Muscles move the eyeballs about and our neck turns so we can see on each side of our bodies. The size of the opening (pupils) in our eyes is changed to control the amount of light that enters the eye. Muscles also change the thickness and curvature of the lens to focus. The image of what we see is upside down on the retina of the eye and the brain interprets these images as right side up. We have two eyes for depth perception. Closing our eyelids allows us to sleep and blinking wipes the surface of the eyes clean, like windshield wipers. Tears provide the washing fluid and ducts drain the tears from the eyes to the nose. Eyelashes act as a protective screen and warn the eyes to close if dust strikes them. Eyebrows and facial bones form a protective wall, with the eyebrows also keeping sweat out of the eyes.

   Darwin acknowledged: "To suppose that the eye with all its inimitable contrivances for adjusting the focus to different distances, for admitting different amounts of light, and for the correction of spherical and chromatic aberration, could have been formed by natural selection, seems, I freely confess, absurd in the highest degree."[2] There is a big difference between "5% of an eye" and "5% vision." For the eye to function at all it must have all of its complex parts working together.³ An eye only partially evolved would be totally useless. The camera was patterned after the eye and has been called "a man-made eye." If the camera was made, doesn't it seem reasonable to believe that the eye also had a Maker?⁴

2. *The Ear.* The ear is probably the second most important sense organ, in that it enables us to hear. It also gives us a sense of balance. The ear consists of several parts:

A. The auricle is the outer ear, which is cuplike to collect sound waves. It also gives us a sense of direction and protects the eardrum from getting damaged.

B. The external auditory canal guides sound waves to the inner ear and is lined with fine hairs and tiny glands that produce wax. These hairs and wax trap dust, insects and other particles, to keep them from causing injury deeper in the ear. The wax also seals the ear for swimming.⁵

C. The middle ear consists of an eardrum and three bones (hammer, anvil and stirrup) with the hammer vibrating the eardrum. The Eustachian tube allows air from the nose to the ears to equalize pressure on both sides of the eardrum.

D. The inner ear consists of semicircular canals that give us our sense of balance. The organ of Corti enables us to hear.

E. The normal ear can hear sounds with the power of only one 10,000 trillionth of a watt. This is so little energy that, if our ears were only slightly more sensitive, we could hear molecules in the air colliding. This is what we hear when we put a shell to our ear and hear the "sound of the ocean." Tiny skeletal muscles dampen vibration.⁶ The Bible says, "The hearing ear and the seeing eye, The LORD has made them both" (Proverbs 20:12).

3. *The Nose.* Our nose, which is used for breathing and for smelling, has been called the world's finest air conditioning plant.⁷ Some parts of our nose are as follows:

A. Two nostrils are separated by a septum.

B. The nasal passages have a lining of soft, moist mucous membrane covered with microscopic, hairlike projections called cilia that wave back and forth to move dust and bacteria to the throat for swallowing.

C. Turbinate bones warm the air before entering the lungs.

D. End fibers of the olfactory nerve carry sensations of smell to the brain.

4. *The Tongue.* Man's chief organ of taste is a two-ounce organ called the tongue. The tongue also helps us to chew and swallow and clears out our

mouth. The taste buds on the tongue enable us to distinguish between sweet, sour, bitter, and salty tastes. Our tongue also helps form sounds like "d, h, j, 1, n, s, t, w, and z." The tongue is sensitive to temperature and it is better to burn your tongue, rather than the throat or stomach. The tongue is also an indicator of sickness.[8]

5. *The Skin*. The skin is the largest organ of the body. It enables us to respond to sensations of pain, texture, heat and cold. The fingertips are the most sensitive to touch or pressure, while the skin of the hands and face have fewer nerve endings that detect heat or cold. The outer layer of skin is called the epidermis. The middle layer, which is called the dermis, contains blood vessels, nerves, sweat and oil glands, and hair follicles. The underskin contains tissue that insulates, stores energy, and cushions muscles, bones, and body organs.

While our body is relatively hairless compared to animals, the hair on our bodies serves as a dry lubricant and scent catcher. Our skin is far more sensitive than the skin of animals.[9] To convey information about temperature and other body conditions to the brain, the skin alone has about four million structures that are sensitive to pain. Pain is caused by nerve endings and keeps us from burning or freezing. The body has a remarkable and complex system to keep its temperature at about 98.6 °F. Heat is produced by burning food. The body dissipates heat through blood flow to bare skin or through approximately two million sweat glands.[10] The evaporation of sweat causes cooling.

What are the major systems of the body?

1. The skeletal-muscular system forms the framework of the body. A newborn baby has 350 bones in its body that must gradually fuse together into 206 adult bones.[11] These bones, along with 650 muscles, tendons and ligaments, enable the body to move. For example, in order to catch a ball, 200 muscles must move in a split second. The bones manufacture red blood cells. The human body has amazing skill to balance and has incredible strength. It was documented that Maxwell Rogers lifted the end of a 3,600-pound car, when the jack holding it up collapsed and the car fell on his son.[12]

The skeletal-muscular system consists of the following:

A. The head contains twenty-two bones and is the control center for the body. It is the protective casing for the brain and eyes. The upper half, called the cranium, contains eight bones, houses our brain and serves as a control tower. The lower part is our face and has fourteen bones. The face has holes for the eyes, ears and nose. The jaw enables us to eat, sing and talk. The facial muscles enable us to squint, smile, frown, purse our lips, or widen our eyes to express surprise, pleasure, disgust, love, fear, and more. It is possible for the human face to make more than 250,000 changes.[13]

B. The spine consists of twenty-four bones called vertebrae. There are soft discs between each vertebrae which stop the bones from hitting each other. The nerve system from the brain travels through the spine.

C. The forearm has two bones allowing the wrist to twist and the hand to turn. The elbow functions like a pulley.

D. The hand which has twenty-seven bones (eight in the wrist, nineteen in the fingers), has been called nature's best tool. It is better than any metal or wooden tool. It can be a cup for water or a hook for a bucket. It can operate a hammer, 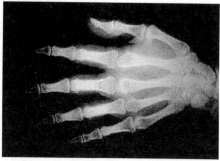 tweezers, scissors, hairbrush and many other devices. It also can help one feel items to make decisions, such as when dirt is too wet or an apple is too soft. The thumb alone is a marvelous item. Without it one would have a difficult time doing things such as writing, picking up coins, or throwing balls.

E. The ribs provide a cage for your heart, liver and lungs. They have an amazing ability to mend when broken.

F. The kneecap stops the leg from bending forward, and the knees allow a person to run, jump, and kick. Cartilage keeps bones from rubbing against each other.

G. The arches in your foot act like shock absorbers and springs. This is necessary for the protection of your spine or backbone. The arches also allow us to run, jump and change directions quickly.[14]

2. *The digestive system* prepares food so the body can use it and eliminates non-used substances. The thirty-two teeth (including four wisdom teeth) are the hardest material in the body, breaking and tearing food into small pieces. Saliva in the mouth moistens food and starts digestion by the use of enzymes.[15] Water is very important for the digestive and circulatory systems. About 62% of our weight is water. Our mouth and throat get dry and tell us we need water.[16] The stomach stores food and produces enzymes that digest the food. The stomach must dissolve food and yet not dissolve itself. The acid in the stomach would eat the varnish off a kitchen

table in seconds.[17] The pancreas and small intestine also produce enzymes. The liver produces bile, which helps the intestines absorb fats and get rid of anything in food that could make us sick. The liver also stores vitamins and iron. If the liver quits, we are dead within eight to twenty-four hours.[18] Cells in the walls of the small intestine absorb food that the blood carries to all parts of the body. The large intestine absorbs water and other liquids, and the rectum moves the solid waste out of the body.

3. *The urinary system* removes urea and other waste products from the body in liquid urine. The blood carries the wastes to the kidneys which, in turn, remove waste products and water from the blood. The liquid passes through the ureter to the bladder. Urine is stored in the bladder until it passes from the body through the urethra.[19]

4. *The respiratory system* brings air into the body and removes carbon dioxide through the process of breathing. Oxygen is needed to break down food and release energy. We breathe air through the nose. The spaces behind the nose warm, moisten, and filter the air. It reaches the lungs where oxygen enters the blood and carbon dioxide leaves it. Carbon dioxide passes out of the body through the nose. Carbon dioxide forms as a waste product when foods are broken down. The body works by burning, and burning food requires oxygen. The respiratory system works because the muscles of the chest and diaphragm contract to enlarge the chest. When the lungs are pulled out by the enlarged chest, they suck in air. Sneezing gets rid of a large speck of pollen or dust in the nose. Coughing gets rid of irritants in the throat. Words are formed by the vocal cords producing a wide range of sounds that are modified by the tongue, teeth, lips, and movements of the cheek. The human voice can produce notes ranging from the lowest Great E^b to the highest $C^{\#}4$. The unaided human voice has been heard as far as six miles away.[20] The protruding chin of the human allows room for considerably more freedom of the tongue; thus we can make far more sounds than any animal.[21]

5. *The circulatory system* includes the heart, a one pound organ, which pumps blood through the arteries to all parts of the body. The blood carries food, oxygen, and waste products to and from the cells.[22] Lack of blood flow to any part of the body can cause damage, especially to the brain. The veins return the blood from the body to the heart. Blood makes over six hundred complete trips through the body each day. The heart beats on the average of seventy times each minute (over 100,000 times a day) to move the blood 168 million miles around the body. If my calculations are correct, the heart beats over two billion times in a life span of seventy years. What pump designed by man can boast this kind of record without an overhaul, much less grow to meet expanding needs? The blood contains platelets and fibrin, which form a quick bandage when cuts

occur. The blood also contains antibodies to fight diseases. The adrenal glands are set at the top of our kidneys. Adrenaline gives an added boost of energy. Women have picked up the ends of cars which have fallen on a husband or child. The pancreas gives sugar to the blood, which gives us energy.

6. *The nervous system* consists of three parts:

   A. The central nervous system includes the brain and spinal cord.

   B. The peripheral nervous system connects the brain and spinal cord with various parts of the body, including the eyes, nose, ears, lungs, heart and digestive systems.

   C. The automatic nervous system regulates the activity of many organs that are not controlled by the conscious mind, such as the stomach, bladder, heart, adrenal glands and sweat glands.

The human brain weighs just three pounds and is carefully packed in the case called the skull. It rests in watery fluid to protect it from serious bumps and shocks. It is the most complex arrangement of matter anywhere in the universe. It can be compared to a video camera, a library with a capacity estimated at 25 million volumes, a computer which handles the information of 1,000 super computers, and a modern national communications center interconnected by billions of telephone wires all in one. There are over 10 billion neurons in the brain, each of which is in dendritic contact with 10,000 other neurons. The total number of neurological interconnections is equivalent to all the leaves on all the trees of a vast forest covering half of the United States.[23] Although there is very little understood about the brain, it is felt that different parts of the brain control different functions.

- The cerebrum is considered the thinking part, controlling speech, hearing, and memory.

- The cerebellum commands our muscles and controls coordination.

- The medulla controls body organs, such as the heart, lungs, and stomach.[24]

The whole body system functions as a unified whole to enable us to talk, run, remember, create, and many other phenomenal tasks we usually take for granted. Darwin thought that the brain of a savage would be only slightly superior to that of an ape. When it was discovered that the brain of a savage was only very little inferior to that of an average member of society, Darwin feared the death of his evolutionary theory. The brain truly provides an ultimate design challenge for evolution. It is a privilege to dedicate these minds to the Creator. We can control our thoughts (Philippians 4:8). Let the words of our mouths and the meditation of our hearts be acceptable in God's sight (Psalm 19:14).

7. *The reproductive system* enables life to continue. The ovaries of the female produce egg cells, which each weigh about one-millionth of a gram. The testes of the male produce sperm, which each weigh about five billionth of a gram.[25] The sperm unites with the egg to start the reproductive process. From this tiny beginning, all the sense organs and systems develop. The fertilized egg then enters the uterus of the female, attaches to the wall of the womb and develops into a baby. "You made all the delicate, inner parts of my body, and knit them together in my mother's womb...My bones were not hidden from Thee when I was made in secrecy and intricately fashioned in utter seclusion" (Psalm 139:13-15 The New Berkley Version). The coded instructions of the original fertilized egg are copied and contained in every cell of the body. The umbilical cord is the source of all food and also the channel of waste for the baby. Living cells are made up of complex protein molecules. Science has only begun to understand the machine-like working of a cell. To propose that a living, replicating cell arose without design from non-living matter is probably the weakest point of evolutionary theory. Thus, from the smallest atom to the entire universe, one can see the design of the Creator. Those who refuse to see this are willingly ignorant.

I want to emphasize that the organs and systems of the human body did not evolve but were *made* by our Creator.

How does a mother's milk give evidence for design?

The milk from the human breast contains all the infant's needs regarding vitamins, minerals, acids, enzymes, carbohydrates, etc. The design of the mother's milk surpasses that of any formula, even milk from cows or goats. What is amazing is that the concentration of the milk changes as the baby ages. For example, the newborn needs more nitrogen and protein, but these concentrations decline as the baby grows older. What if even one of these essential ingredients was missing? "If the mother's milk was not well designed so perfectly from the beginning, humankind would be extinct." There is no way that random chance could have produced such an awesome product.[26]

Why do manufacturers of most items provide written operating instructions?

An instruction manual comes with almost every new electrical or mechanical device that is purchased. These booklets may go by other names, such as "Safety Instruction and Operator's Manual," "Use and Care Instruction Manual," "Use and Care Guide," "Operating Instructions," or "Owner's Manual." The manufacturers, who have designed and manufactured the product, write the manuals, and they should know better than anyone else how to get the best results from the electrical or mechanical device.

Consider the following examples:

1. *Garage door opener.* Caution: Read instructions and rules for safe operation carefully. Fasten this manual near the garage door after installation. Periodic checks of opener are required to insure satisfactory operation. Failure to comply with the following instructions may result in serious personal injury or property damage.

2. *Electric Can Opener.* When using electrical appliances, basic safety precautions should always be followed, including reading all instructions before using.

3. *Coleman Lantern.* Follow instructions and warning to avoid injury or fires.

4. *Vacuum Cleaner.* Before you vacuum, read the owner's manual. Follow instructions while you're vacuuming. Use your cleaner only for the job it was intended to do.

5. *Dual Stereo Cassette Player/Recorder with AM/FM Stereo.* Know your unit. Read this booklet so you will be able to enjoy all the features.

6. *Lawnmower.* Read this manual thoroughly. If you don't understand any portion, contact your dealer for a demonstration of actual starting and operating procedures.

Note from the above, the possible consequences if you don't follow the manufacturer's instructions.

1. Serious personal injury.

2. Property damage.

3. Fires.

4. Lack of enjoyment of the features.

5. Ruining the product.

We are cautioned to read the instructions *before* we operate the product. Sadly for some of us, we take the philosophy "When all else fails, read the instructions."

What book contains our Creator's instructions?

If we are fearfully and wonderfully made, does our Manufacturer (Creator) have any instructions? Yes! The Word of God!

What are the consequences if you don't follow His instructions?

1. Personal injury and damage. The way of the transgressor is hard. Many pitfalls can be avoided by knowing and obeying God's Word.
2. Lack of enjoyment of His blessings.
3. Eternal damnation.

I want to encourage you to read God's Word and follow it. Get to know the Creator. He knows best how you should operate your life.

[1] Pavlu, 146.

[2] Charles Darwin, *The Origin of the Species*, A. L. Burt Co., London 1859.

[3] Johnson, 34.

[4] Jackson, *The Human Body - Accident or Design?* 57.

[5] William Coleman, *My Magnificent Machine*, Bethany Fellowship, Inc., Minneapolis, MN 55438, 1978.

[6] Larry Vardiman, "The Human Ear and How it Interprets Sound," Science, Scripture and Salvation Radio Broadcast, Program No. 475 aired weekend of 6/17/95.

[7] Jackson, *The Human Body - Accident or Design?* 65.

[8] Coleman.

[9] Mark P. Cosgrove, *The Amazing Body Human, God's Design for Personhood*, Baker Book House, Grand Rapids, MI 49516, 1987, 55.

[10] Jerry Bergman, "Mankind - The Pinnacle of God's Creation," Impact Article No. 133, Institution for Creation Research, El Cajon, CA, 92021, 1984.

[11] Jackson, *The Human Body - Accident or Design?* 19.

[12] Bergman, Impact Article No. 133.

[13] Cosgrove, 30-31.

[14] Coleman.

[15] Jackson, *The Human Body - Accident or Design?* 33-35.

[16] Coleman.

[17] Bergman, Impact Article No. 133.

[18] Coleman.

[19] Jackson, *The Human Body - Accident or Design?* 68-72.

[20] Bergman, Impact Article No. 133.

[21] Cosgrove, 41.

[22] Jackson, *The Human Body - Accident or Design?* 38-39.

[23] Donald B. DeYoung, and Richard B. Bliss, "Thinking About The Brain," Impact Article No. 200, Institution for Creation Research, El Cajon, CA, 1990.

[24] Jackson, *The Human Body - Accident or Design?* 50.

[25] Cosgrove, 26.

[26] Rex. D. Russell, "Design in Infant Nutrition," Impact Article No. 259, Institute for Creation Research, El Cajon, CA, January 1995.

35 / Radiometric Dating Methods

How is radioactivity used to determine the age of a sample?

To understand radioactive dating, a person must have some knowledge of chemistry, which is the study of the structure and properties of matter. Matter is made up of elements (pure substances, of which there are slightly over one hundred), compounds (chemical combinations of the elements) and mixtures of the above. The atomic theory describes the atom as the smallest particle of an element that still has the property of that element. An atom consists of electrons and a nucleus made up of protons and neutrons. For example, an ordinary carbon atom has six protons, six neutrons and six electrons. It is known as carbon-12 because the sum of the protons and neutrons equals twelve. Elements can have isotopes, which are atoms with the same number of protons but a different number of neutrons. Carbon, for example, has another isotope known as carbon-14 because it has six protons and eight neutrons. A radioactive isotope is one that is not stable and decays by giving off some form of radiation. Carbon-14 is a radioactive isotope.

The radiocarbon dating method utilizes the fact that, in the upper atmosphere, cosmic rays react with nitrogen to form carbon-14. Also, in the upper atmosphere, carbon reacts with oxygen to form carbon dioxide. Thus, some carbon dioxide contains carbon-12 and some contains carbon-14. Plants and indirectly animals ingest this radioactive carbon dioxide along with normal carbon dioxide. Carbon-14 decays to carbon-12. The time for which half of the carbon-14 decomposes to carbon-12 is known as the half-life of carbon-14, which is estimated to be 5,760 (plus or minus 30) years. The longer the plant or animal has been dead, the less will be the ratio of carbon-14 to carbon-12. If one analyzes the amount of radioactive substance in a sample and compares it to the amount of radioactive substance it originally was assumed to have, one can attempt to calculate the age of the substance using the half-life.[1]

What assumptions are made in radiometric dating?

1. One must assume knowledge of the original amount of both the radioactive isotope and the decay product. The ratio of carbon-14 to carbon-12 for carbon dioxide is not yet in equilibrium. Who knows what the ratio of carbon-14 to carbon-12 was thousands of years ago? The cosmic ray influx would have had to remain constant in order to estimate the amount of carbon-14 originally present. In the uranium-lead dating method, it is assumed that there was no lead in the original sample.

2. Changes in these concentrations can only occur by radioactive decay, i.e., none of these changes can occur by leaching or any other process but

123

radioactive decay. Could not some of the substance be washed in or out by the liquids in the soil or decomposed due to other factors? As much as 90% of the total radioactive elements of some granites could be removed by leaching the granulated rock with weak acid.[2]

3. The rate of decay is constant regardless of external conditions. Who knows that the rate of decay has remained constant over thousands of years? Could temperature, pressure and other factors have affected this rate?

What are some major errors in dates that have resulted from radiometric dating?

Carbon dating often arrives at dates much different from those determined by archaeological studies or other methods. For example:

1. Shells of snails living in southern Nevada gave an apparent age of 27,000 years.

2. A shell from a live clam was dated thousands of years old.

3. Dried seal carcasses less than 30 years old were dated as old as 4,600 years, and a freshly-killed seal was dated at 1,300 years old.

The potassium-argon radiometric dating method was used on some volcanic rocks off Hualalai in the Pacific Islands of Hawaii. These rocks were erroneously dated as being 160 million and 3 billion years old. Actually, these rocks were formed during a volcanic eruption in the year 1801.[3] Also, radiometric dating methods used on volcanic rock on Reunion Island in the Indian Ocean gave results varying from 100,000 to 4.4 billion years.[4]

Thus we can conclude that dates obtained by radiometric methods are interesting geo-physical exercises, but they do not provide accurate information as far as the age of the earth is concerned.[5]

---

[1] Michell J. Sienko and Robert A. Plane, Chemistry 2nd Edition, McGraw-Hill Book Company, Inc., New York, NY, 1961, 573-574.

[2] Taylor, *Illustrated Origins Answer Book*, 61.

[3] Morris, *Scientific Creationism*, 147.

[4] Taylor, *Illustrated Origins Answer Book*, 61-62.

[5] Morris, *Biblical Basis*, 260-269.

124

36 / Depletion of the Earth's Magnetic Field

How could the decaying of the earth's magnetic field give evidence of a young earth?

The three important force fields associated with planet earth are as follows:

1. The gravitational field attracts us to the earth and keeps us from flying off.

2. The electric field is very unstable and produces electrical storms.

3. The magnetic field turns compasses. It is thought that a huge electric current circulating in the core of the earth causes this field. The basic field is called the dipole field in that it has two fields - a north and south pole.

The earth's magnetic field is decaying faster than any other worldwide geophysical phenomenon. In 1835 Karl Gauss made the first evaluation of the earth's magnetic dipole moment (the vector which gives the strength and direction of the magnet). Evaluations have been made every 10 or 15 years since then. The decay seems to be exponential with a half-life of 1400 years.[1] This means it is cut in half every 1400 years. Or extrapolating backwards, it was doubled every 1400 years. This, of course, assumes a constant rate of decay, which has not been actually observed prior to 1835. The following graph visually plots this assumed relative strength of the earth's magnetic field over time.

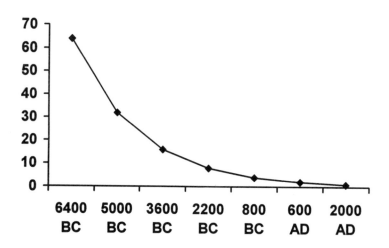

If one assumes that the magnetic force has always been decaying at a constant rate, and if one would extrapolate backwards, say 20,000 years, one would find that the earth could not have survived the heat generated by stronger currents. This means there is a limit to the age of the earth, because there is a limit as to how much magnetic energy the earth could have originally had.[2] The magnetic field would have been so strong that if a man was chopping wood, the axe would rotate toward the magnetic pole, and sailing ships with metal could only sail north or south. The idea of a young earth is disastrous to evolutionists because their whole theory depends on eons of time. They therefore postulate that the magnetic poles of the earth have reversed every so often. For this they have absolutely no evidence.[3]

---

[1] Thomas G. Barnes, "Depletion of the Earth's Magnetic Field," Impact Article No. 100, Institute for Creation Research, El Cajon, CA, October 1981.

[2] Ibid.

[3] Thomas G. Barnes, "Earth's Magnetic Age -The Achilles Heel of Evolution," Impact Article No. 122, Institute for Creation Research, El Cajon, CA, August 1983.

37 / Dendrochronology

How can the age of a tree be determined?

Dendrochronology (dendron = tree, chronos = time, logos = word = the study of) is the field of study that uses the annual growth rings of trees to establish dates.[1] Fritt broadly defined dendrochronology to include all tree-ring studies where the annual growth layers have been assigned to, or are assumed to be associated with, specific calendar years.[2] Webster defines dendrochronology as the science of dating events and variations in the environment in former periods by comparative study of growth rings in trees and aged woods.[3] Trees can not be dated merely by their height or diameter because the growth rate of trees depends on many factors, such as slope gradient, sunlight, wind, soil properties, temperature, moisture, rainfall, floods, snow accumulation, glacial activity, atmospheric pressure, volcanic activity, disease, pests and even variations in nearby stream flows. Thus two trees of the same species, in the same area, can be of the same age yet vary greatly in size. The way to estimate age is by using the tree rings.

What causes tree rings?

In most non-tropical areas of the world, tree rings are formed by the layer of wood that is added each year to the trunk and branches of a tree. New wood grows between the old wood and the bark. In the spring, when moisture is plentiful, trees grow rapidly. In the summer, growth progressively slows down until the fall when growth stops. No new growth appears until the next spring. The contrast between these periods of rapid and slow growth establishes a ring that makes counting possible.[4] Tree rings usually extend around the entire circumference of the tree. Normally, the rings are annual rings, but in lower elevations or latitudes a tree could produce two or more growth rings in one year. In the tropics, where seasons are not so distinct, trees can grow year-round. Thus most tropical trees do not form annual rings, and there are few, if any, tree-ring sites in the tropics between 30 degrees north and south latitudes.[5]

How can tree ring counts be made on living trees?

A proper tree ring count can only be made at the base of a tree. A horizontal cross section can be obtained either by cutting down a tree or using a stump from a tree that has already been cut down or fallen. Increment borers are used to date living trees by taking a horizontal sample from the tree. The borer is an auger-like instrument with a hollow shaft that is screwed into the trunk, which extracts from the tree a long slender core sample about .423 centimeters in diameter (less than the diameter of a pencil).[6] The borer, however, is not long enough to reach the center of a large tree.[7] After a core has been extracted, the hole should not be plugged since trees are able to repair a small wound by filling the hole with resin, which prevents contamination. Plugging the hole with a stick or anything else could introduce diseases.[8]

Can the oldest trees be used to date the Flood?

Trees never die of old age, but only from injury, disease or by accident. The giant sequoias, commonly known as California redwoods, are very resistant to insects, disease and fire.[9] Although capable of living indefinitely, the greatest age of the giant sequoia is approximately 4,000 years old. Some bristle cone pine trees located in the White Mountains of California have also been found with very old ages. The earth's oldest living inhabitant seems to be a bristle cone pine tree called "Methuselah" estimated to be 4,763 years old.[10] One of these trees found in the Snake Range of Nevada has actually been estimated to be 4,900 years old. However, due to the possibility of multiple rings per year and the fact that the rings are exceedingly thin, none seem to be much older than 4,000-4,500 years old.[11] It is interesting that this would be the approximate date of the Flood in Noah's day. It appears that the great flood must have destroyed almost all trees at that time, since fossils have been found of these tree types. Many of the trees buried during this time probably turned to coal. Seeds have survived flooding for long periods of time and probably replanted most of these trees after the Flood.[12]

How is the crossdating technique used to try to date dead trees or wood samples?

Crossdating utilizes the fact that the size of tree rings can vary greatly because of the varying growth rate of the trees. When the growth of a tree has been limited by environmental factors, such as slope gradient, poor soils and little moisture, more variation in ring to ring growth will be present. This variation is referred to as sensitivity, and the lack of ring variability is called complacency. This generally occurs with constant climatic conditions such as a high water table, good soil or protected locations.[13] Crossdating is the technique of matching a series of rings from one tree with a series from another tree by analyzing the unique patterns of narrow and wide rings. For crossdating to work, there must be a full range of rings from very narrow to very wide, in order to see common patterns. The trick is to date it against a master reference that has been previously anchored in time.[14]

What about the claims of 8,000 year-old trees by some dendrochronologists?

Some dendrochronologists claim that trees go back at least 8,000 years without being disturbed by Noah's flood. This date, however, was not arrived at by counting the tree rings of just one living or dead tree, but rather required the superimposing of tree rings of about twenty different dead trees.[15, 16] This would mean that samples of trees would have to be used that had been dead for over 3,000 years. Even though the bristlecone pine is a resinous tree, I find it hard to believe that with weathering, decay, termites, etc., these dead tree samples would last for over 3,000 years in an unprotected state.

The bristlecones are not large trees. The tallest is, at most, about sixty feet and the girth of the largest one, the Patriarch, is less than thirty-seven feet. The girth is the distance around the tree. For the bristlecone, this is usually not a circle as in many trees, because the tree is snarly and twisted. Each year the girth of the tree increases only about 1/100 of an inch. The rings are very thin, sometimes only a few hundredths of a millimeter each, and thus patterns are sometimes very hard to recognize from tree to tree.[17] The rings can only be seen with a high powered microscope. Another complicating factor is that some bristlecones have multiple stems from the same root mass, with some stems being much older than others.[18]

Even trees on the same mountain and of the same species can have considerably different tree-ring patterns. A fire can affect the growth rate of a tree. It can increase the growth rate of a tree, if nearby plants that had been competing for nutrients and water have burned off. A fire can decrease the rate of growth, if many of its leaves are burned off, because the tree has lost some of its ability to photosynthesize (which occurs in the foliage), or create nutrients from sunlight and water.

Considering the fact that samples from multiple trees were used, the rotting problem, the smallness of the rings, and the many factors that could cause variation in size and shapes of rings, many dates achieved by crossdating are questionable.

Why do evolutionary dendrochronologists try to achieve older dates for their wood samples?

"The desire to stretch out any age-dating calculation is apparently practically universal among evolutionary geochronologists. The older things can be made to appear, the less reliable the Bible seems to be, and the further back God is pushed from any meaningful connection with the world He created."[19] Several quotes from Miller show this to be true. "The hope is to push the date back to at least 8,000 BC. This will be important as the last Ice Age ended about 10,000 years ago, and to have a record of this transition period would offer scientists a wealth of information."[20] The desire is to get dates that match the evolutionary assumption that the ice age ended about 10,000 years ago. He also wrote that "The bristlecone chronologies had raised questions regarding radiocarbon dating methods and have been used to recalibrate the C-14 process."[21] This reminds me of a story I once heard of a man whose job it was to shoot off a cannon every morning at 6:00 A.M. to start the day at the army base. On the way to the base, he purposely passed a jewelry store to set his watch by the clock in the window. One day he needed his watch repaired, so he brought it to the jewelry store. While there, he inquired as to how they knew that their clock in the window was accurate. They said that they set it each morning at 6:00 A.M. when the cannon sounded. Can you imagine the accuracy of C-14 dating if it is calibrated by the crossdating of dendrochronologists? This is circular reasoning at its worst.

Another disturbing item is that the Uniformitarian Principle, "the present is the key to the past," originally stated by James Hutton in 1785, is now considered to be one of the basic principles of dendrochronology. In other words, physical and biological processes that link current environmental processes with current patterns of tree growth must have been in operation in the past.[22] This would rule out catastrophes like the Flood in Noah's day, yet at the same time they want to correlate the dating with the ice age, which also would have been a catastrophe with no similar situation occurring in the present. This seems like a contradiction to me.

130

[1] Henri D. Grissino-Mayer, "Principles of Dendrochronology," The Tree Ring Web Pages, http:/tree.ltrr.arizona.edu/~grissino/princip.htm, 1997.

[2] Ibid.

[3] *Webster's Ninth New Collegiate Dictionary*, 1990 edition.

[4] Leonard Miller, The Ancient Bristlecone Home Page, http://www.sonic.net/bristlecone/intro.html, 1995-1997.

[5] Henri D. Grissino-Mayer, "Photo Gallery of Trees and Tree Rings," The Tree Ring Web Pages, http:/tree.ltrr.arizona.edu/~grissino/gallery.htm#History, 1997.

[6] Grissino-Mayer, "Principles of Dendrochronology."

[7] Grissino-Mayer, Personal correspondence, December 14, 1997.

[8] Grissino-Mayer, "Photo Gallery of Trees and Tree Rings."

[9] Whitcomb and Morris, *Genesis Flood*, 392-393.

[10] Miller, Ancient Bristlecone.

[11] Morris, *Biblical Basis*, 449-450.

[12] Dorothy E. Kriess Robbins, "Can the Redwoods Date the Flood?" Impact Article No. 134, Institute for Creation Research, El Cajon, CA, August 1984.

[13] Miller, Ancient Bristlecone.

[14] Grissino-Mayer, Personal Correspondence, December 14, 1997.

[15] Morris, *Biblical Basis*, 451.

[16] Morris, *Scientific Creationism*, 163, 193.

[17] Morris, *Biblical Basis*, 450.

[18] Grissino-Mayer, Personal correspondence, December 12, 1997.

[19] Ibid., 452.

[20] Miller, Ancient Bristlecone.

[21] Ibid.

[22] Grissino-Mayer, "Principles of Dendrochronology."

38 / Other Dating Methods

How does the influx of materials into the ocean give evidence for a young earth?

Both dissolved and suspended materials are carried into the oceans by the rivers. Uniformitarians believe that present day rates and processes are similar to those in the past, "...all things continue as they were from the beginning of the creation" (II Peter 3:4). Several methods could then be used to determine the age of the earth, if one assumes uniform rates, which is a very questionable assumption.

River deltas. If a river delta, such as at the base of the Mississippi River, increases on an average of one mile every sixteen years, and if the earth is approximately 6,000 years old, then the delta would be about 375 miles long, which isn't impossible.[1] However, if the earth were billions of years old, the oceans would have been filled long ago.

Sediments at the bottom of the ocean. It is estimated that 27.5 billion tons of sediment are being transported to the ocean every year. If this rate had continued for 4.5 billion years, the sediment at the bottom of the ocean would be much deeper than it is today.

Dissolved chemicals. "The amount of any given chemical in the ocean, divided by the annual increase of that chemical through river inflow, will yield the time required to accumulate the chemical, assuming none was present in the ocean to begin with, and the inflow rate has always been the same."[2] The most common chemicals in ocean water are the elements that make up salt (sodium and chlorine). The corresponding chlorine calculation determines the age of the ocean as ninety million years. Both of these are far less than five billion years. Also, these calculations assume no salt in the original ocean. Couldn't God have created a salt-water ocean as well as a fresh-water ocean? It seems likely, if the sea creatures were to be salt-water fish, that God would have made a salt-water ocean for them. If you assume a salt concentration to begin with, the age calculated is much less, depending on what concentration you assume. If one uses aluminum concentrations in the ocean, one would determine the age of the earth as one hundred years old.[3] We know the earth is much older than that. Thus, this uniformitarian method of aging is a very questionable method. Obviously, in the past, aluminum wasn't being poured into the oceans at the same rate as it is today.

How does the lack of helium in the earth's atmosphere give evidence for a young earth?

The earth's atmosphere is made up of approximately 78% nitrogen, 21% oxygen, and traces of other gases, including helium. While some of this helium could have come from the original Creation and some from the sun, most was probably produced in the earth's crust by radioactivity that gives off alpha particles, which are the nuclei of helium atoms. This helium would then seep through the surface of the earth and subsequently mix in the atmosphere.[4] If the rate of this helium flow from the earth has been constant and the earth is billions of years old, there should be up to a million times more helium in the earth's atmosphere than now exists. Evidence indicates that very little helium escapes into space.[5] The fact that there is so little helium in the earth's atmosphere gives evidence for a young earth.

How does the lack of dust on the moon show that the earth is quite young?

Evolutionists generally assume that the age of the moon is the same as the age of the earth. Space dust, which probably results from the breakup of comets, meteors and meteorites, regularly settles on the surface of the earth and the moon. It has been estimated that over 14 million tons of dust settles on the surface of the earth each year. Much more is burned up in the earth's atmosphere. Whereas the moon has less surface area and gravitational attraction, the moon does not have a substantial atmosphere to burn up the space dust.[6] On the earth some of this dust could be mixed with the soil and some washed into the seas by wind and water. On the moon though, there is no wind or rain resulting in little or no erosion. The moon then should have an increasing depth of dust accumulation. Assuming a constant rate of deposition and that the moon was 4.5 billion years old, it was thought that the dust on the surface of the moon would be over a hundred feet deep. Because of this assumption, and the fear that the lunar landing module would sink completely out of sight when it touched the moon, the lunar landing modules were designed with huge saucer-shaped feet.[7] To the surprise of the astronauts on Apollo 11, the dust on the moon was only two to four inches deep.[8] At a constant rate of deposition this would determine the age of the moon (and the earth) to be quite young, perhaps less than 10,000 years old.

When the astronauts planted the United States flag on the moon, they planned to stick the pole into the lunar surface. However, it wouldn't go in deep enough because the dust layer was too thin. In order to support the flag in an upright position they had to rig up a telescoping device. They were very concerned about the possibility of the American flag collapsing in front of the television cameras.[9]

---

[1] The World Book Encyclopedia, 1975, Vol. 5, 102.

[2] Morris, *Scientific Creationism*, 153.

[3] Harold Camping, "Let the Oceans Speak," Creation Research Society Quarterly, Vol. 11, June 1974, 42.

[4] Larry Vardiman, "Up, Up, and Away! The Helium Escape Problem," Impact Article No.143, Institute for Creation Research, El Cajon, CA, May 1985.

[5] Taylor, *Illustrated Origins Answer Book*, 14, 65, 66.

[6] Ibid., 69, 70.

[7] Ibid., 17.

[8] DeYoung, *Astronomy and the Bible*, 32-34.

[9] http://www.msnbc.com./.onair/msnbc/TimeAndAgain/archive/Apollo/mission.asp

39 / Theistic Evolution

How much difference is there between biblical and evolutionary concepts of time?

In chapter 4, I pointed out that the biblical record shows that man was created about 4000 B.C. Thus man was created about 6,000 years ago. The earth and universe were created only a few days before man. Jesus confirms this by saying, "But from the beginning of the creation God made them male and female" (Mark 10:6). Thus Jesus put the creation of man in the same time frame as the creation of the earth. On the other hand, evolutionary cosmologists say that the earth and the solar system developed about five billion years ago, that primitive life began evolving about four billion years ago and finally man evolved one or two million years ago. Thus the biblical chronology is about a million times shorter than evolutionary chronology. This is a tremendous difference. Can the pre-human chronology of Genesis 1 be reconciled with a five billion year earth history, and can the pre-Flood and post-Flood human chronologies of Genesis chapters 5 and 11 be reconciled with a one or two million year human history?[1]

Why have some tried to harmonize the Genesis account of Creation with evolutionary theories?

Early Bible commentators accepted the literal six days of Creation. It has only been since the middle of the nineteenth century that commentators began talking about long periods of time within Genesis 1 itself. It was at this time that Darwin's *Origin of Species*, Lyell's *Principles of Geology*, and other evolutionary treatises were flooding the marketplace, resulting in a widespread popular acceptance of the major tenets of evolution. Unfortunately, many theologians thought geologists had positive evidence that the world was exceedingly old. Instead of holding their ground and insisting on the authenticity of God's account of origins, many theologians made evolutionary theory the criterion of truth. They felt it was necessary to compromise the biblical account of origins with the speculations of nineteenth-century atheists and agnostics.

What is "theistic evolution?"

By "theistic evolution," we mean that evolution from gases to man was directed by God, rather than governed by the rules of chance and natural selection as believed by evolutionists.[2] In other words, God used the mechanism of evolution to create the world and all that is in it. To some, this may seem on the surface to be acceptable, but in reality, belief in a creator God and belief in evolution are totally opposite viewpoints. Julian Huxley, an evolutionist, said "God is unnecessary," and G.G. Simpson referred to church services as "higher superstitions celebrated weekly in every hamlet of the United States."[3] I don't believe a Christian can be an evolutionist, once one understands what the Bible teaches and what evolution really is. The term "theistic evolutionist" is about as consistent as "Christian thief " or "righteous murderer."

It is very sad that Christian people have been so quick to compromise with the atheistic philosophy of evolution. Many theologians wrote about evolution as God's "method of Creation," either forgetting or not knowing that it was all supposed to be accomplished by a brutal struggle for existence, with only the fittest surviving and the weakest perishing.

What are some examples of the devastating effect of evolution on faith?

Charles Darwin is an example of the impact of evolutionary belief on Christian faith. As a young man, Darwin prepared for the Christian ministry. As he came to believe in evolution and natural selection, he increasingly lost his faith and finally became an atheist. Darwin's notes prove that he had become an atheist some twenty years before he published *The Origin of Species by Natural Selection.* He adopted natural selection, rather than the hand of God, as the active factor responsible for all that was formerly considered evidence for design.

The decline and fall of Darwin's faith has been repeated in the experiences of multitudes of others since his day. One of the top modern-day evolutionists claims to have been a born-again Christian in high school but left this fundamental religion when he got to the University of Alabama and heard about evolutionary theory. Probably more cases of lost religious faith can be traced to the theory of evolution than to anything else.

Henry Morris wrote, "In spite of this record, however, there are multitudes of professing Christian people who think they can believe both the Bible and evolution -- that evolution is merely God's method of creation. One can only say that anyone who believes this simply does not understand either evolution or the Bible or both."[4]

What are some reasons why evolution cannot be harmonized with the biblical record of Creation?

1. The phrase "after his kind" is used ten times in Genesis 1:11, 12, 21, 24, 25. While the biblical "kind" is probably more flexible than the biological "species," this restriction limits all variation to variation within the kind. Changing from one kind to another is not possible (I Corinthians 15:38, 39).

2. At the end of the Creation week, "God ended his work," and "rested from all his work" which He created and made (Genesis 2:2-3). Thus present-day biological processes, variation, mutation and even speciation, could not be processes of creation or development, as theistic evolutionists must assert.

3. At the end of the six days of Creation, God declared that all His work of Creation was "very good" (Genesis 1:31). It would be inconsistent for an omniscient, loving God to say that a system of nature ruled by tooth and claw, where only the fittest or more prolific survive, would be "very good."

4. The Lord Jesus Christ, who is the Creator of all things (Colossians 1:16), taught that the Genesis record of Creation was historically and literally accurate. (See Matthew 19:4-6, Mark 10:6-9.) He stated that "from the beginning of the creation God made them male and female" (Mark 10:6), not from the tail end of four billion years of evolutionary history. Also, God made man and woman to have dominion over His Creation (Genesis 1:26, 28). This command to have dominion would have been irrelevant for most of the Creation if animals had been created millions of years before this time.

5. I agree with Dr. Henry Morris, who wrote, "Evolution is the most wasteful, inefficient, and heartless process that could ever be devised by which to produce man. If evolution is true, then billions upon billions of animals, have suffered and died in a cruel struggle for existence for a billion years, and many entire kinds (e.g. dinosaurs) have appeared and then died out long before man evolved. The God of the Bible could never be guilty of such a cruel and pointless charade as this!"[5]

---

[1] Morris, *Biblical Basis*, 115.

[2] Paul A. Zimmerman, "Can We Accept Theistic Evolution?" A Symposium on Creation, Baker Book House, Grand Rapids, MI, 1968, 56.

[3] Ibid., 76.

[4] Morris, *Biblical Basis*, 111-113.

[5] Ibid., 113-114.

40 / The Day-Age Theory

What are some ways theologians attempt to reinterpret biblical chronology?

The Day-Age Theory considers the "days" of Genesis 1 to be comparable to the "ages" of geology, thus putting the "geologic ages" during the six days of Creation.

The Pre-Adamic Gap Theory inserts the geologic ages between Genesis 1:1 and 1:2 and before the six days of Creation.

The Post-Adamic Gap Theory assumes gaps in the genealogical lists of Genesis 5 and 11, thus allowing a human history of more than six thousand years.[1]

My first exposure to one of these interpretations was a variation of the Day-Age Theory. This view taught the following: There were different kinds of days mentioned in the Bible: Solar days were twenty-four hours in duration, dispensational days were 1,000 years long and creative days were 7,000 years each. We are living in the sixth day, which is man's day. The sixth day will be 7,000 years, based on the eschatological viewpoint that the Old Testament was 4,000 years. The New Testament will be 2,000 years and the Millenium will be 1,000 years. Every other creative day was also 7,000 years because the term "the evening and morning were" was used for each of these days. Since the sixth day will end in judgment, each of the other days also ended in judgment. According to this theory, coal was formed because of the judgment at the end of the third day. The judgment at the end of the fifth day caused the dinosaurs to become extinct and oil to form. Thus, according to this interpretation, the earth is about 41,000 years old. The seventh day will be eternal because there is no mention of "evening and morning were" at the end of the seventh day and the seventh day has not yet come.

There are several problems, however, with this interpretation. The Bible doesn't say we are living in the sixth day, that each day was 7,000 years or of equal length, or that there was judgment at the end of each day. The term "rested" in Genesis 2:2 is in the past tense, thus it has already occurred. Also, 41,000 years is not enough to satisfy the evolutionists, thus most Day-Age theorists propose very long days.

What is the alleged basis for the Day-Age Theory?

In Genesis 1, the Hebrew word for "day" is *yom*. This word is used in a variety of ways to indicate either the daylight portion in the normal day, a normal 24-hour period or an indefinite time period as referenced in Psalm 90:10. A passage that is often referred to is II Peter 3:8, "One day is with the

Lord as a thousand years and a thousand years as one day." Some also claim that too much activity took place on the sixth day (Genesis 2) to fit into a normal day. These activities include the naming of thousands of animals, Adam's perception of loneliness and the creation of Eve.[2] The claim is that the days of Genesis 1 are really long periods of time which correspond to the major periods of evolutionary geological history.

How can the Day-Age Theory be refuted?

1. *An improper interpretation of II Peter 3:8.* II Peter 3:3-10 is a unit. The context speaks of scoffers in the last days who will ridicule the Second Coming of Christ. Their logic is that Jesus promised to come quickly but He has not yet come, therefore He is not going to come at all. They say, "...all things continue as they were from the beginning of the creation." Peter refutes these uniformitarian assumptions with a reference to the Flood and the certainty of judgment for these scoffers. Peter wrote these words in response to the charge that Christ failed to fulfill His promise.

   II Peter 3:8 is not a mathematical formula of 1 = 1000 or 1000 = 1. "As" is not the same as an equal sign. God created time and therefore stands above it. While we think 1,000 years is a long time, God doesn't. The verse could have been worded, "One minute with the Lord is as ten thousand years," and still have conveyed the same message. God was able to do in one day what would normally require a thousand years to accomplish. Thus one day's flood activity could build up layers of sediments that would normally take a thousand years to form by natural processes. The context of II Peter 3:8 has nothing to do with the length of the Creation week, but rather deals with the Flood. Genesis 1 needs to be interpreted by its own context.

2. *The inadequacy of a thousand-year day.* If each of the six periods were of equal length, then six days would result in a 6,000 year period of creation, but 6,000 years is not adequate to make the evolutionary system work. A young earth deals a death blow to the evolutionary theory. If one day equals 7,000 years, then six days would equal 42,000 years, which is not enough time either. If one day equals one billion years that would fit the evolutionary time scale. However, if words have this kind of infinite flexibility the art of communication is a lost cause. Moses could have used the Hebrew word *olom* if he had intended to convey the idea of a long time period. Instead he used the Hebrew word *yom* to convey the idea of a 24-hour day.

3. *The demands of primary word usage.* The primary usage of any word is always given priority in a translation, and secondary uses are tried only when the primary usage does not make sense in the context. The Hebrew word *yom* is used more than 2,000 times in the Old Testament. In over 95% of the cases the word clearly means a 24-hour day or the daylight

portion of a normal day. Thus, an unbiased translator would normally understand the idea of "24-hour period" for the word *yom* even without a context. Many of the other 5% refer to expressions such as "the day of the Lord" (Joel 2: 1).

4. *The numerical qualifier demands a 24-hour day.* The word "day" appears over two hundred times in the Old Testament with numbers (i.e., first day, second day, etc.). In each case it refers to a 24-hour day. Since the six days of the creation week are so qualified, consistency of usage requires a 24-hour day in Genesis 1 as well.

5. *The terms "evening and morning" require a 24-hour day.* The words evening and morning always refer to normal days where they are used elsewhere in the Old Testament. The Jewish day began in the evening at sunset and ended with the start of the evening on the following day.

6. *The words "day" and "night" are part of a normal 24-hour day.* Nine times in Genesis 1: 5, 14-18, the words day and night are used to indicate the light and dark periods of a normal 24-hour day.

7. *Genesis 1:14 distinguishes between days, years, and seasons.* "And God said, Let there be lights in the firmament of the heaven to divide the day from the night; and let them be for signs, and for seasons, and for days, and years." The words days, years, and seasons mean what they say. Some claim that if the sun did not appear until the fourth day there could be no days and nights on the first three days. However, all that was needed was a light source and a rotating earth for there to be alternating periods of light and darkness (evenings and mornings) for the first three days (Genesis 1:3-13).

8. *Symbiosis requires a 24-hour day.* Symbiosis is a biological term describing a mutually beneficial relationship between two types of organisms. Many plants cannot reproduce apart from the habits of certain insects or birds. For example, most flowers require bees or other insects for pollination and reproduction. Plants were created on the third day, birds on the fifth day and insects on the sixth day (Genesis 1:11-25). While plants could have survived for 48 or 72 hours without the birds and the bees could they have survived two to three billion years without each other, according to the day-age scenario? Could the birds have survived a billion years while waiting for the insects to evolve for food?

9. *The survival of the plants and animals requires a 24-hour day.* If each day were a billion years, then half of that day (500 million years), would have been dark. The light was called day and the darkness was called night, and each day had one period of light-darkness. How would the plants, insects, and animals have survived through each 500 million year time of darkness? Obviously a 24-hour day is indicated.

140

10. *The testimony of the Fourth Commandment.* Exodus 20:8-11 reads as follows: "Remember the Sabbath day, to keep it holy. Six days you shall labor and do all your work, but the seventh day is the Sabbath (rest) of the LORD your God...For in six days the LORD made the heavens and the earth, the sea, and all that is in them, and rested the seventh day."

Verses 8-10 speak of man working six days and ceasing from his work on the seventh. These are normal 24-hour days. Moses equated the time period of creation with the time period of man's workweek (six days plus one day) and indicated that the Lord had set the example in Genesis 1. The same author, Moses, wrote both Genesis 1 and Exodus 20 at about the same time, and had the same time period in mind when he used the word day. The Fourth Commandment was actually written by the finger of God. Surely the Creator knew how long the days were.

11. *The theological problem of sin and death.* According to evolutionists, plant and animal life flourished and died at least 500 million years before man evolved. The Bible, however, teaches that there was no death prior to Adam's sin in the Garden of Eden. (See Genesis 3, Romans 5:12-14.) Either the Day-Age Theory is wrong, or my understanding of the Bible is in error. One must make a choice.[3] The Day-Age Theory is not a satisfactory compromise.

---

[1] Morris, *Biblical Basis,* 116.

[2] William J. Spear, Jr., "Could Adam Really Name All Those Animals?" Impact Article No. 165, Institute for Creation Research, El Cajon, CA, July 1995.

[3] Richard Niessen, "Theistic Evolution and the Day-Age Theory," Impact Article No. 81, Institute for Creation Research, El Cajon, CA, March 1980.

41 / The Gap Theory

What is the Gap Theory and why did it originate?

The Gap Theory was another attempt by theologians to reconcile the apparent short scale of world history found in Genesis with the popular belief that geologists had positive evidence that the world is very old (dinosaurs, coal, the Grand Canyon, etc.). Thomas Chalmers (1780-1847), a Scottish theologian and the first Moderator of the Free Church of Scotland, probably was the person most responsible for the origin of the Gap Theory.[1] Currently this idea, also known as the ruin-reconstruction view, is held by many who use the Schofield Reference Bible or the Dake's Annotated Bible.[2]

This theory can be summarized as follows:

In the beginning God created a perfect heaven and earth. This original creation of the universe was followed by billions of years of evolutionary development. "Satan was ruler over the earth which was peopled by a race of men without any souls. Eventually Satan, who dwelled in a garden of Eden composed of minerals (Ezekiel 28) rebelled by desiring to become like God (Isaiah 14) Because of Satan's fall, sin entered the universe and God judged the earth by means of a flood (Genesis 1:2) and then a global ice-age where light and heat were somehow removed. All the plant, animal, and human fossils upon the earth today date from this "Lucifer's flood" and don't bear any relationship to plants and animals upon the earth today."[3]

They would translate the first part of Genesis 1:2 to read "and the earth became without form and void" or "But the earth became ruined and empty." Thus they believe a large "gap" of time existed between Genesis 1:1 and Genesis 1:2 sufficient enough to account for the geologic ages evidenced by the so-called geologic column. Gap theorists (or "gappists") would claim to believe in a literal view of Genesis but oppose a recent origin of the universe. They would propose that God reshaped the earth and re-created all life in the six literal days after Lucifer's Flood, which produced the fossils, hence the name "ruin-reconstruction."

One of the arguments for the Gap Theory is that the Hebrew words *bara* and *asah* always have different meanings. They allege that *asah* cannot mean "to create." However, there are a number of verses where the word *asah* does mean create such as Nehemiah 9:6, "God made *asah* the heaven, even the highest heavens, and all their starry host, the earth and all that is on it, the seas and all that is in them" (NIV). In actuality, these two words are used interchangeably in certain contexts.[4]

Many gappists advocate that the grammar of Genesis 1:1-2 allows, and even requires, a gap of time between what is recorded in verse 1 and verse 2.[5]

This is not evident by reading the text. Dr. Raymond Crownover heard me speak on this subject several years ago and wrote to me saying:

> "In Hebrews there is a grammatical construction known as the Wou Consecutive. This construction links two or more verbs by appending the Hebrew letter Wou to the second and succeeding verbs. In the King James Version this construction is translated "and" but, although it is a connective it also presents the idea of consecutive action. That is each of the verbs connected by the Wou Consecutive are understood to take place consecutively during the same time period. Both verses 2 and 3 of Genesis chapter 1 begin with the Wou Consecutive construction tying these verses to verse 1 as occurring during the same period without any gap in time. The Hebrew text of Genesis 1, therefore disallows any gaps at least between the "original creation" of verse 1 and the first creative day and night of verses 2 and 3."[6]

This grammatical connection between verses 1 and 2 rules out the Gap Theory because verse 2 is a description of the originally created earth. The New International Version reads "Now the earth was formless and empty."

Another argument gappists use is in the King James Version of Genesis 1:28 where it states "...Be fruitful and multiply and replenish the earth ..." In reality this did not mean restock to the English readers of that day, but accurately reflected the original Hebrew, which means simply "fill." It cannot be used to say that God meant to refill the earth.[7]

What are some biblical arguments against the Gap Theory?

1. There is no scriptural evidence that Lucifer's fall produced a flood and there is no mention of Lucifer's flood anywhere in the Bible.[8]

2. The idea that fossils formed before the "reconstruction" of the earth contradicts the fact that there was no death before Adam's sin (Romans 5:12).[9]

3. This theory proposes that men existed before Adam, while the Bible states "The first man Adam..." (I Corinthians 15:45).[10]

4. If fossils were produced by Lucifer's flood then what effect did Noah's Flood have? Noah's Flood would have removed all traces of the fossil record, unless it was a local flood,[11] which has been shown in chapter 6 to be unscriptural.

5. Could God have summarized everything He made as very good (Genesis 1:31) if Lucifer had already fallen and there was death and destruction due to Lucifer's flood?

6. Exodus 20:11 states, "For in six days the Lord made the heavens and the earth, the sea, and ALL THAT IN THEM IS (emphasis mine)..." Where is there time for a gap?[12]

7. The Gap Theory does not resolve the problem of evolution but merely pushes it back into a pre-Genesis world. This theory implies that God used evolutionary methods in the pre-world, and then changed to direct creative activity in the six days of "re-creation".

---

[1] Ken Ham, Andrew Snelling, and Carl Wieland, *The Answers Book - Answers to the 12 Most-asked Questions on Genesis and Creation/Evolution*, Master Books, El Cajon, CA, 1992, 157.

[2] Ibid., 174.

[3] Weston W. Fields, *Unformed and Unfilled*, Presbyterian and Reformed, Nutley, NJ, 1976, 7.

[4] Ham, Snelling and Wieland, 166-167.

[5] Ibid., 170.

[6] Raymond Crownover, Personal correspondence.

[7] Ken Ham, "Closing the Gap," Back to Genesis Article, Institute for Creation Research, El Cajon, CA, February 1990.

[8] Morris, *Biblical Basis*, 121-122.

[9] Pavlu, 170.

[10] Morris, *Biblical Basis*, 122-123.

[11] Ham, Snelling and Wieland, 161.

[12] Ibid., 162.

42 / Six Literal 24-Hour Days of Creation

First of all, I don't think a person is lost who does not believe in six literal 24-hour days of creation. I was saved and believed a form of the Day-Age Theory because of what I was taught. Later, I was exposed to and embraced the Gap Theory because it seemed to be a better explanation for the ideas commonly taught in geology classes. Now, after realizing that there is no scientific evidence whatsoever for evolution or the geologic ages, I believe in six literal 24-hour days of creation.

Why I believe in six literal 24-hour days of creation.

1. A proper interpretation of the Bible as discussed in chapters 39, 40 and 41, shows that the six days of creation were of 24-hour duration.

2. Since the theory of evolution is not scientific fact one does not need ideas such as theistic evolution, the Day-Age Theory, or the Gap Theory to harmonize the Bible with the theory of evolution.

3. God is a God of miracles. When God said, "Let there be light" (Genesis 1:3), there was light, not just a glimmer, glow or ray. When I turn on the light switch for an incandescent bulb, there is light immediately. Some of the key words in Mark are immediately and straightway. Expressions, such as "immediately the fever left," "immediately the leprosy departed," "immediately he arose," "straightway his ears were opened," and "immediately he received his sight," are common in the gospel according to Mark. With God nothing is impossible. The big question is not "How could God have created the earth and all that is in it in six days?" but rather "Why did it take Him so long?" I believe that God could instantaneously form light, the firmament, dry land, vegetation, the sun, the moon, the stars, the birds, the fish and the animals. Because of this, I have no trouble believing that God can instantaneously forgive and deliver a man from sin and fill him with His spirit. I also believe that I shall be changed in a moment, in a twinkling of an eye and be caught up to meet the Lord in the air. (See I Thessalonians 4:13-18, I Corinthians 15:51-54.) Have I ever seen this take place? No. Did Noah ever see rain before the Flood? No. "By faith Noah, being divinely warned of things not yet seen, moved with godly fear, prepared an ark for the saving of his household..." (Hebrews 11:7).

Are creationists or evolutionists biased?

Most creationists have the bias that the biblical record of Creation is true and view all things in light of the Scriptures. On the other hand, most evolutionists are biased to believe in the theory of evolution with its concept of geological ages and they interpret all things from this viewpoint. Both are biased. "It's not a matter of whether you are biased or not, but which bias is the best bias to be biased with anyway. The bias that you have determines what you do with the evidence when you find it. This applies to scientists as well."[1]

---

[1] Ken Ham, "The Relevance of Creation," Taped from Institute for Creation Research, El Cajon, CA.

43 / Evolutionary Religion

Why do most people believe in the theory of evolution?

1. Most people believe in the theory of evolution because they have been taught, or brainwashed, that it is scientific fact which has been established by supposedly intelligent, honest scientists.[1] From the time a child learns to read, he is presented with storybooks of dinosaurs and cavemen with an evolutionary slant. They are taught by teachers who believe in evolution.

2. Most liberal theologians, and even some conservative theologians, have felt it necessary to bend their interpretation to fit what they consider scientific truth.[2]

3. Major universities, scientific publishing houses and periodicals are under the control of evolutionists. Many won't print a serious scientific article or book refuting evolution or even a letter to the editor.[3] I have tried several times. At some universities it is almost impossible for a Creationist to have an influential position on the faculty in disciplines dominated by evolutionary philosophies.[4]

4. The average person has never had the opportunity to hear intelligent, informed people declare that evolution is a theory, which is not only unsupported, but contradictory to established scientific laws.

5. Also, the museums and the media bombard the minds of people with evolutionary theory presented as fact.

It is no wonder that the average person believes in evolution rather than the biblical truth. They need to hear the truth from someone like you. For faith comes by hearing the Word of God (Romans 10:17).

Is evolution a religion?

Both Creation and evolution are religious views. The issue is not science vs. religion, but religion vs. religion.[5] I have previously shown that any concept regarding origins is not scientific, in that origins were not and cannot be observed, repeated or verified. The scientists can only deal with present evidence. The choice of which theory to accept becomes a matter of faith.[6] To accept something without evidence requires faith. Hebrews 11:1-3 states, "Now faith is the substance of things hoped for, the evidence of things not seen...By faith we understand that the worlds were framed by the word of God, so that the things which are seen were not made of things which are visible." The Christian believes that God created the universe, life and man, while the evolutionist believes that the universe, life and man somehow evolved without any supernatural direction. "Evolution cannot be proved or tested, it can only be believed."[7] Considering the majesty, beauty and

147

complexity of the earth and universe, it is relatively easy to believe in Creation. But to believe that dead matter could create life, and have absolutely no evidence, requires faith of another order.[8] Some believe that a cosmic egg of energy exploded to form chemical elements, stars, galaxies and finally people. Some even have the faith to believe that life was planted on earth by an unknown civilization from outer space.[9] Since evolution cannot be observed, repeated or verified, it is no more "scientific" and no less "religious" than Creation.[10] One person was asked, "Why aren't you an evolutionist?" His reply was, "I don't have enough faith to believe that random particles arranged themselves into ordered life."

The zeal of Darwinists to evangelize the world with their theory makes it also seem like a religion. They see evolution as a light which illuminates all facts. To them evolution is the god they worship.[11]

Why do some people, after being confronted with the concept of God and creation, still wish to believe in the theory of evolution?

1. Some accept evolution because they are afraid of what others would think if they didn't.

   As evidence, let me take some direct quotes from a brochure entitled "Keeping Creationism Out of Education: What Parents and Teachers Can Do" distributed by the National Center for Science Education, Inc., PO Box 9477 Berkeley, CA 94709.

   - "When colleagues make it clear to a creationist teacher that his/her behavior is viewed unfavorably by colleagues, this peer pressure is often sufficient to discourage it."

   - "Point out to the teacher that the scientific establishment is firmly on the side of evolution."

   - "The job of the teacher is to present the consensus of the discipline." (In other words, continue to perpetuate ideas that may not be true if it is the consensus of the discipline.)

   - "Although the teacher may "believe" creationism is scientifically respectable, no one else does and it is his/her responsibility to present state-of-the-art science."

2. For others, the alternative to evolution is the awareness that there may well be a personal Creator to whom one must give account. Many people wish to live immoral lives without the thought of being judged by God someday. To believe in evolution is to disbelieve the God of the Bible.

At least two portions of Scripture use the illustration of the potter and the clay. In Jeremiah 18:1-7 the potter is God and the clay is Israel. God showed that he could pluck up, pull down, or destroy nations as easily as a potter can form vessels from clay.

In Romans 9:20-23, Paul again shows that the potter has power over the clay. He can make vessels of honor or vessels of dishonor fit for destruction.

Men realize that if God created them, He has a right to punish them for their sins just as a potter has a right to destroy vessels that don't turn out right. Men try to push these thoughts out of their minds by believing in evolution.

Can a Christian be an evolutionist?

To answer this question one must define terms. I will define a Christian as one who believes and obeys the teachings of Jesus Christ. Jesus believed in Creation (Matthew 19:4), and the Flood (Matthew 24:37, 38). In fact, "...by him were all things created that are in heaven, and that are in earth..." (Colossians 1:16). Thus a Christian will believe in the Genesis account of Creation as well as in the rest of the Bible. In contrast to this, consider these quotations:

- Julian Huxley said, "In the evolutionary pattern of thought there is no longer either need or room for the supernatural. The earth was not created, it evolved. So did all the animals and plants that inhabit it, including our human selves, mind and soul as well as brain and body."[12]

- Cornell University Professor William Provine, a leading historian of science, insists in order for people to retain religious belief and accept evolutionary biology they "have to check their brains at the church-house door."[13]

- Richard Dawkins wrote, "In Darwin's view the whole point of the theory of evolution by natural selection was that it provided a non-miraculous account of the existence of complex adaptations."[14]

- Douglas Futuyama stated "...the human species was not designed, has no purpose, and is the product of mere mechanical mechanisms."[15]

In light of the above quotations it can be concluded that an evolutionist does not believe in special Creation as recorded by the Bible but explains origins by natural means only. Thus a Christian cannot be an evolutionist unless he does not understand what Jesus taught or what evolution means. The theory of evolution contradicts what the Bible teaches.

An evolutionist, however, can become a Christian, when he becomes aware of God as his Creator and turns to Jesus Christ as his Savior.

---

[1] Morris, *Twilight of Evolution*, 26, 93.

[2] Ibid., 20-21.

[3] Ibid., 27.

[4] Ibid., 27-28.

[5] Ham, *The Lie*, 16.

[6] Pavlu, 104.

[7] Henry M. Morris, "Evolution is Religion not Science," Impact Article No. 107, Institute for Creation Research, El Cajon, CA, May 1982.

[8] Henry M. Morris, "The Splendid Faith of the Evolutionists," Impact Article No. 111, Institute for Creation Research, El Cajon, CA, September 1982.

[9] Ibid.

[10] Walter G. Peter, III, "Fundamentalist Scientists Oppose Darwinian Evolution," Bioscience, October 1, 1970, 1067.

[11] Johnson, 9, 132.

[12] Ibid., 153-205.

[13] Ibid., 126, 202.

[14] Ibid., 178.

[15] Ibid., 202.

References

1. Anderson, Larry, "Oil made from Garbage," Science Digest, Vol. 74, July 1973.
2. "Arctic Tern," Microsoft® Encarta® 96 Encyclopedia, ©1993-1995 Microsoft Corporation.
3. Austin, Steven A., "Springs of the Ocean," Impact Article No. 98, Institute for Creation Research, El Cajon, CA, August 1981.
4. Austin, Steven A., "Ten Misconceptions About The Geologic Column," Impact Article No. 137, Institute for Creation Research, El Cajon, CA., November 1984.
5. Austin, Steven A., "Mount St. Helens and Catastrophism," Impact Article No. 157, Institute for Creation Research, El Cajon, CA, July 1986.
6. Austin, Steven A., "How Fast Can Coal Form?" Creation Ex Nihilo, Vol. 12, No. 1, December 1989-February 1990.
7. Baker, Sylvia, Bone of Contention – Is Evolution True? Creation Science Foundation Ltd., Queensland, Australia, 1986.
8. Barnes, Thomas G., "Depletion of the Earth's Magnetic Field," Impact Article No. 100, Institute for Creation Research, El Cajon, CA, October 1981.
9. Barnes, Thomas G., "Earth's Magnetic Age -The Achilles Heel of Evolution," Impact Article No. 122, Institute for Creation Research, El Cajon, CA, August 1983.
10. Basic Science Physics Pace #135, Reform Publications, Inc., 1987.
11. Bergman, Jerry, "Mankind - The Pinnacle of God's Creation," Impact Article No. 133, Institution for Creation Research, El Cajon, CA, July 1984.
12. Bergman, Jerry, "The Earth: Unique in the Universe," Impact Article No. 144, Institution for Creation Research, El Cajon, CA, June 1985.
13. Bliss, Richard B., Parker, Gary E., Gish, Duane T., Fossils: Keys to the Present, Creation Life Publishers, PO Box 15908, San Diego, CA, 1980.
14. Bowden, Malcolm, Ape-Men: Fact or Fallacy, second edition, Bromley, Kent, England, Sovereign Publications,1981.
15. Camping, Harold, "Let the Oceans Speak," Creation Research Society Quarterly, Vol. 11, June 1974.
16. Chapman, Geoff, "The Giraffe," Creation Ex Nihilo, Volume 12, No.1, December 1989-February 1990.
17. Chemical and Engineering News, May 29, 1972.
18. Coffin, Harold G., with Brown, Robert H., Origin By Design, Review and Herald Publishing Association, Washington, DC, 1983.
19. Coleman, William, My Magnificent Machine, Bethany Fellowship, Inc., Minneapolis, MN, 1978.

20. Cosgrove, Mark P., *The Amazing Body Human, God's Design for Personhood*, Baker Book House, Grand Rapids, MI, 1987.

21. Crownover, Personal correspondence.

22. Darwin, Charles, *The Origin of the Species*, A. L. Burt Co., London 1859.

23. DeYoung, Donald, B., "The Moon: A Faithful Witness in the Sky," Impact Article No. 68, Institute for Creation Research, El Cajon, CA, February 1979.

24. DeYoung, Donald B., "Design in Nature, The Anthropic Principle," Impact Article No. 149, Institute for Creation Research, El Cajon, CA, November 1985.

25. DeYoung, Donald B., *Astronomy and the Bible, Questions and Answers*, Baker Book House, Grand Rapids, MI, 1990.

26. DeYoung, Donald B., and Bliss, Richard, "Thinking About The Brain," Impact Article No. 200, Institution for Creation Research, El Cajon, CA, February 1990.

27. DeYoung, Donald B., *Weather and the Bible, 100 Questions and Answers*, Baker Book House, Grand Rapids, MI, 1993.

28. Eldredge, Niles, as quoted in: Luther D. Sunderland, *Darwin's Enigma: Fossils and Other Problems*, fourth edition (revised and expanded), Master Book Publishers, Santee, CA,1988.

29. Fields, Weston W., *Unformed and Unfilled*, Presbyterian and Reformed, Nutley, NJ, 1976.

30. Gish, Duane T., "The Amino Acid Racemization Dating Method," Impact Article No. 23, Institute for Creation Research, El Cajon, CA, 1975.

31. Gish, Duane T., "The Origin of Mammals," Impact Article No. 87, Institute for Creation Research, El Cajon, CA, September 1980.

32. Gish, Duane T., "The Mammal-like Reptiles," Impact Article No. 102, Institute for Creation Research, El Cajon, CA, December 1981.

33. Gish, Duane T., "Evolution and the Human Tail," Impact Article No. 117, Institute for Creation Research, El Cajon, CA, March 1983.

34. Gish, Duane T., *Evolution, The Challenge of the Fossil Record*, Creation Life Publishers, El Cajon, CA, 1985.

35. Gish, Duane T., "Remarkable Scientific Accuracy of Scripture," Institute for Creation Research, El Cajon, CA, Science, Scripture and Salvation; Weekly Broadcast No. 165 aired weekend of July 8, 1989.

36. Gish, Duane T., "Startling Discoveries Support Creation," Impact Article No. 171, Institute for Creation Research, El Cajon, CA, September 1987.

37. Gish, Duane T., "Modern Scientific Discoveries Verify the Scriptures," Impact Article No. 219, Institute for Creation Research, El Cajon, CA, September 1991.

38. Gish, Duane T., "Petroleum in Minutes Coal in Hours," Vol. 1, No. 4.

39. Gitt, Werner, "The Flight of Migratory Birds," Impact Article No. 159, Institute for Creation Research, El Cajon, CA, September 1986.

40. Grantham, Brian, "My Favorite Evidence for Creation," Creation Ex Nihilo, Vol. 12, No. 1, December 1989-February 1990.

41. Grissino-Mayer, Henri D., the "Principles of Dendrochronology," The Tree Ring Web Pages, http:/tree.ltrr.arizona.edu/~grissino/princip.htm, 1997.

42. Grissino-Mayer, Henri D., "Photo Gallery of Trees and Tree Rings," The Tree Ring Web Pages, http:/tree.ltrr.arizona.edu/~grissino/ gallery.htm#History, 1997.

43. Ham, Ken, *The Lie - Evolution*, Master Books, P.O. Box 1606, El Cajon, CA, 1987.

44. Ham, Ken; Snelling, Andrew; and Wieland, Carl, *The Answers Book - Answers to the 12 Most-asked Questions on Genesis and Creation/Evolution*, Master Books, El Cajon, CA, 1992.

45. Ham, Ken, "Closing the Gap," Back to Genesis Article, Institute for Creation Research, El Cajon, CA, February 1990.

46. Ken Ham, "The Relevance of Creation," Taped from Institute for Creation Research, El Cajon, CA.

47. Ham, Ken, "Watches and Wombats," Back to Genesis Article, Institute for Creation Research, El Cajon, CA, March 1990.

48. Ham, Ken, "The Smartest Man in America?" Back to Genesis Article No. 48, Institute for Creation Research, El Cajon, CA, December 1992.

49. Ham, Ken, "Dinosaurmania Strikes Again," Back to Genesis No. 55, Institute for Creation Research, El Cajon, CA, July 1993.

50. Hand, John Raymond, "Why I Accept the Genesis Record," Back to the Bible Broadcast, Box 233, Lincoln, NE, 1959.

51. Hastie, Peter, "What Happened to the Horse?" Creation Magazine, Sep.-Nov. 1995, Vol. 17, No. 4, pp. 14-16, supplied by Answers in Genesis, http://www.christiananswers.net/q-aig/aig-c016.html

52. Humber, Paul G., "The Ascent of Racism," Impact Article No. 164, Institute for Creation Research, El Cajon, CA, February 1987.

53. Hunt, Kathleen, "Horse Evolution," The Talk, Origins Archive, Exploring the Creation/ Evolution Controversy, January 4, 1995.

54. Jackson, Wayne, *Creation, Evolution, and the Age of the Earth*, Courier Publications, P.O. Box 55265, Stockton, CA, 1989.

55. Jackson, Wayne; *The Human Body, Accident or Design?* Courier Publications, P.O. 55265, Stockton, CA, 1993.

56. Johnson, Phillip E., *Darwin on Trial*, InterVarsity Press, Downers Grove, IL, 1993.

57. Klotz, John W., *Genes, Genesis, and Evolution*, Concordia Publishing House, St. Louis, MO, 1955.

58. Lewin, Roger, *Bones of Contention: Controversies in the Search for Human Origins*, Simon & Schuster, 1987.

59. Long, Michael E., "Secrets of Animal Navigation," National Geographic, June 1991.

60. Meldau, Fred John, *Why We Believe in Creation Not Evolution,* Christian Victory Publishing Co. , Denver CO, 1959.

61. Miller, Leonard, The Ancient Bristlecone Home Page, http://www.sonic.net/bristlecone/intro.html, 1995-1997.

62. Morris, Henry M., *The Twilight of Evolution,* Baker Book House, Grand Rapids, MI, 1963.

63. Morris, Henry M., *Scientific Creationism* (Public School Edition), CLP Publishers, San Diego, CA, 1974.

64. Morris, Henry M., "Probability and Order Versus Evolution," Impact Article No. 73, Institute for Creation Research, El Cajon, CA, July 1979.

65. Morris, Henry M.; *Men of Science - Men of God,* Master Books, A division of Creation Life Publishers, P.O. Box 1606, El Cajon, CA, 1982.

66. Morris, Henry M., "Bible Believing Scientists of the Past," Impact Article No. 103, Institute for Creation Research, El Cajon, CA, January 1982.

67. Morris, Henry M., *The Biblical Basis for Modern Science,* Baker Book House, Grand Rapids, MI, 1984.

68. Morris, Henry M., "The Heritage of the Recapitulation Theory," Impact Article No. 183, Institute for Creation Research, El Cajon, CA, September 1988.

69. Morris, Henry M., "Dragons in Paradise," Impact Article No. 241, Institute for Creation Research, El Cajon, CA, July 1993.

70. Morris, John D., "Mount St. Helens: Explosive Evidence for Creation," Notes presented at Institute of Creation Research Summer Institute, Northwestern College, St., Paul, MN, July 11, 1988.

71. Morris, John D., "Scripture and The Flood," Notes presented at Institute of Creation Research Summer Institute, Northwestern College, St., Paul, MN, July 11, 1988.

72. Morris, John D., "Was "Lucy" an Ape-man?" Back To Genesis, Institute for Creation Research, El Cajon, CA, November 1989

73. Morris, John D., "Dinosaurs," Science, Scripture and Salvation Broadcast No. 201, Institute for Creation Research, El Cajon, CA, March 17, 1990.

74. Morris, John D., "Did a Watchmaker Make the Watch?" Back to Genesis Article, Institute for Creation Research, El Cajon, CA, March 1990.

75. Morris, John D., "Why Don't We Find More Human Fossils?" Back to Genesis Article No. 37, Institute for Creation Research , El Cajon, CA, January 1992.

76. Morris John D., "What about the Horse Series?" Back to Genesis No. 63, Institute for Creation Research, El Cajon, CA, March 1994.

77. Morris, John D., "What are Polystrate Fossils?" Back to Genesis Article No. 81, Institute for Creation Research, El Cajon, CA, September 1995.
78. Nevins, Stuart E., "The Origin of Coal," Impact Article No. 41, Institute for Creation Research, El Cajon, CA, November 1976.
79. Niessen, Richard, "Theistic Evolution and the Day-Age Theory," Impact Article No. 81, Institute for Creation Research, El Cajon, CA, March 1980.
80. Oiler, John W., Jr., "A Theory in Crisis," Impact Article No. 180, Institute for Creation Research, El Cajon, CA, June 1988.
81. Parker, Gary, "The Origin of Life on Earth," Creation Research Society Quarterly, September 1970.
82. Parker, Gary E., "Creation, Mutation and Variation," Impact Article No. 89, Institute for Creation Research, El Cajon, CA, November 1980.
83. Parker, Gary, "Things That Are Made," Impact Article No. 62, Institute for Creation Research, El Cajon, CA, August 1978.
84. Pavlu, Ricki D., *Evolution: When Fact Became Fiction*, Word Aflame Press, Hazelwood, MO, 1986.
85. Peter, Walter G., III, "Fundamentalists Scientists Oppose Darwinian Evolution," Bioscience, October 1, 1970.
86. Ranganathan, B. G., *Origins?*, The Banner of Truth Trust, 3 Murrayfield Road, Edinburgh EH12 6EC, P.O. Box 621, Carlisle, PA, 1988.
87. Rajca, John, "Flood Accounts Around the World," Science, Scripture and Salvation Broadcast No. 170, Institute for Creation Research, El Cajon, CA, August 12, 1989.
88. Raup, David M., "Evolution and the Fossil Record," letter in Science 213 (July 17, 1981), see also by the same author "Geology and Creationism," Field Museum Bulletin 54 (March 1983).
89. Reno, Cora, *Evolution, Fact or Theory?* Moody Press, Chicago, IL, 1953.
90. Richards, Lawrence, *It Couldn't Just Happen - Faith Building Evidences for Young People*, Word Publishing, Dallas, TX, 1989.
91. Robbins, Dorothy E. Kreiss, "Can the Redwoods Date the Flood?" Impact Article No. 134, Institute for Creation Research, El Cajon, CA, August 1984.
92. Sienko, Michell J., and Plane, Robert A., Chemistry Second Edition, McGraw-Hill Book Company, New York, NY, 1961.
93. Slusher, Harold S., in Glashouwer, William J. J. and Taylor, Paul S., writers, The Earth, A Young Planet?, Mesa, AZ: Eden Films and Standard Media, 1983 (Creationist Motion Picture).
94. Spear, William J., Jr., "Could Adam Really Name All Those Animals?" Impact Article No. 165, Institute for Creation Research, El Cajon, CA, July 1995.
95. Stambaugh, Jim, Science Scripture and Salvation Tape No. 476, Saturday June 24, 1995.

96. Taylor, Paul S., *The Illustrated Origins Answer Book*, Eden Productions, P.O. Box 41644, Mesa, AZ, 1990.

97. Taylor, Stan, "The World That Perished," A video produced by Films for Christ, Mesa, AZ, 1977.

98. The Associated Press, November 2, 1997.

99. *The Nature of Science*, Prentice-Hall Inc., Englewood Cliffs, NJ, 1993.

100. The Record, the newspaper serving San Joaquin County, Stockton, CA, September 12, 1994.

101. Tillery, Bill W., Physical Science Second Edition, William C. Brown Publishers, Dubuque, IA, 1993.

102. Time Life Books, Early Man, Introductory Volume in the Life Nature Library.

103. Tiner, John Hudson, College Physical Science Pace 50003, Accelerated Christian Education, Inc., 2600 ACE Lane, Lewisville, TX.

104. Thorndike-Barnhart, *Comprehensive Desk Dictionary*, Doubleday and Company, Inc., Garden City, NY, 1955.

105. Unruh, J. Timothy, "The Greater Light to Rule The Day," Impact Article No. 263, Institute for Creation Research, El Cajon, CA, May 1995.

106. Vardiman, Larry, "The Sky Has Fallen," Impact Article No. 128, Institute for Creation Research, El Cajon, CA , February, 1984.

107. Vardiman, Larry, "Up, Up, and Away! The Helium Escape Problem," Impact Article No.143, Institute for Creation Research, El Cajon, CA, May 1985.

108. Vardiman, Larry, "The Human Ear and How it Interprets Sound," Science, Scripture and Salvation Radio Broadcast, Program No. 475 aired Weekend of June 17, 1995.

109. Wald, George, "The Origin of Life," Scientific American.

110. *Webster's New World Dictionary of the American Language*, Second College Edition, The World Publishing Company, New York, NY.

111. *Webster's Ninth New Collegiate Dictionary*, 1990 edition.

112. Whitcomb, John C., Jr., and Morris, Henry M., *The Genesis Flood - The Biblical Record and Its Scientific Implications*, Baker Book House, Grand Rapids, MI, 1961.

113. Williams, Emmett L., and Mulflinger, George, *Physical Science for Christian Schools*, Bob Jones University Press, Greenville, SC, 1974.

114. *The World Book Encyclopedia*, World Book-Childcraft International, Inc., Volumes 1, 2, 6, 9, 12, 13, 21, Chicago, IL, 1979.

115. Zimmerman, Paul A., "Can We Accept Theistic Evolution?" A Symposium on Creation, Baker Book House, Grand Rapids, MI, 1968.

Index

To order copies of books written by Arlo Moehlenpah send the following information.

Ship to: (Please Print)

Name			
Address			
City		State	Zip
Phone	Email		
Creation vs. Evolution: Scientific and Religious Considerations	Quantity	Price (each) $11.95*	$
Teaching with Variety	Quantity	Price (each) $8.99*	$
Master Your Money Or It Will Master You	Quantity	Price (each) $11.95*	$
Postage and handling		$1.25 per book	$
Total amount enclosed			$

*Quantity discounts available

Make checks payable and mail to:
Doing Good Ministries
217 Bayview Way • Chula Vista, CA 91910

For information regarding seminars on:
Creation vs. Evolution
Teacher Training
Personal Finance
Principles of Jesus
Principles from the Epistles

Contact Arlo Moehlenpah at: 619-476-7230
Email Moehlenpah@aol.com

Visit our web site at www.DoingGood.org